AN

ACCOUNT

OF THE

FISHES

FOUND IN

THE RIVER GANGES AND ITS BRANCHES

AN

ACCOUNT

OF THE

FISHES

FOUND IN

THE RIVER GANGES AND ITS BRANCHES

BY

FRANCIS HAMILTON

GYAN

Published by

Gyan Publishing House
5, Ansari Road
Daryaganj, New Delhi-110002
Phone: 011-47034999, 9811692060
E-mail: books@gyanbooks.com

Distribution Network
gyanbooks.com
India, USA, Canada, UK, Australia

ISBN: 978-93-6433-035-0 (HB)
First Published, 1822

2nd Impression 2024

Printed at: Gyan Press, Delhi.

AN ACCOUNT OF THE FISHES FOUND IN THE RIVER GANGES AND ITS BRANCHES
Author: FRANCIS HAMILTON

AN

ACCOUNT

OF THE

FISHES OF THE GANGES.

AN

ACCOUNT

OF THE

FISHES

FOUND IN

THE RIVER GANGES AND ITS BRANCHES.

BY

FRANCIS HAMILTON, (FORMERLY BUCHANAN,) M.D.

FELLOW OF THE ROYAL SOCIETIES OF LONDON AND EDINBURGH;

THE SOCIETY OF ANTIQUARIES, AND OF THE LINNÆAN SOCIETY OF LONDON;

AND OF THE ASIATICK SOCIETY OF CALCUTTA.

WITH A VOLUME OF PLATES IN ROYAL QUARTO.

EDINBURGH:

PRINTED FOR ARCHIBALD CONSTABLE AND COMPANY, EDINBURGH;

AND HURST, ROBINSON, AND CO. 90, CHEAPSIDE, LONDON.

1822.

INTRODUCTION.

In this work I have described the Fishes found in the Ganges and its tributary streams. Notwithstanding the great variety of species here mentioned, I am persuaded, that the list is far from being complete. For the two last years, indeed, of my stay in India, and while engaged in countries remote from the sea, I met with but few species that I had not before described; but, previous to my departure for Europe, on returning to the vicinity of the numerous and large estuaries by which this enormous river enters the sea, I daily met with species which I had never before seen. Besides, it is probable, that, in the Alpine sources of the Ganges and its branches, there are a good many species which are not to be found in the rivers of the plains, nor in the torrents among the puny hills of the south. During a year's residence, however, near the capital of Nepal, I observed few species that I had not previously seen; but fishing is not a favourite art among the people of that country, so that my opportunities of investigation were not great. On the whole, I do not think, that, in the rivers above, where the tide reaches, more than one kind in five has escaped my notice;—while of those found in the estuaries I have not probably described above a half. This is the less to be regretted, as a great proportion of them has probably been described in the valuable work of Dr Russell.

In the arrangement of the fishes, I have followed the method of Linnæus; for, although that of La Cepede is more complete, the distinctions arising from the gill-covers and membranes are often traced with so much difficulty, as to occasion much doubt in practice. In arranging the genera and species, however, I have, for the most part, followed La Cepede. I had finished the manuscript before I saw the valuable system of Cuvier; and the numerous changes which would have been required to adopt it, would have occasioned a trouble, for which the improvements, although considerable, would not, I think, be a sufficient counterbalance.

For the sake of my countrymen residing on the banks of the Ganges, to whom this work should be peculiarly useful, I have employed the English language, although the technical terms necessary to be used in describing natural objects appear more uncouth in a modern dialect than in the Latin tongue. I request the reader to observe, that, in speaking of the form of fishes, I use the terms *high* and *flat* to express their great or small vertical breadth, while their horizontal breadths are expressed by the terms *wide* and *narrow*. These terms are, therefore, applied differently in describing the flat fishes which swim on one side, to what they are in describing all the others, which swim with their bellies toward the bottom. In writing of the back and vent fins, by the terms *long* and *short*, I mean the great or small length which these organs occupy on the fishes' body; while by *high* and *low* I express the great or small length of their rays. Again, in writing of the pectoral, ventral, and tail fins, the terms *long* and *short* are applied to their greater or less extent from their root to their end. In describing all fins, the term *rounded* implies that their upper edge, along the extremities of the rays, forms a curve projecting in the middle; and the term *like a crescent*, implies that this edge forms a curve receding in the middle. The English term *oval* implies the botanical Latin term *ovatus*, and not *ovalis*.

For the sake of those men of science to whom the English language is not familiar, I have annexed a Synoptical Table in Latin, following in this the example of LATHAM, in his excellent work on Birds.

To most of the species that I could not trace in authors, I have given trivial names, derived from some of those used by the natives of India, and altered as little as possibly consistent with avoiding that total barbarity into which LA CEPEDE has so often fallen. In this I shall endeavour to imitate the venerable RUMPHIUS, who, from the most barbarous materials, has given many names greatly more polished than the sesquipedalic compounds of the modern dealers in Greek.

In their proper places I have introduced an account of a few fishes observed in the rivers of the south of India; but to these I have prefixed no numbers in the Synoptical Table. These have only been annexed to the species found in the Ganges.

The words printed in capitals or italics are chiefly intended to assist the reader in finding more readily the objects which are described.

ERRATA.

Page 165, after PIMELODUS RITA, Plate XXIV. *insert* and XXV.

Page 361, line first, *expunge* quod.

Plate I. for MUGIL CORSULA *read* PIMELODUS; the MUGIL CORSULA being delineated Plate IX. fig. 97.

A

DESCRIPTION

OF THE

GANGETIC FISHES.

ORDER FIRST.

FISHES HAVING A CARTILAGINOUS SPINE.

I. Genus.—Raia or Skate.

Fishes of the first order, which, on each side of the body beneath, have five or seven breathing apertures without gill covers or membranes; which have the mouth opening from below; and which have the body very flat, and terminated on the sides by a margin composed of the pectoral and ventral fins.

1st Species.—Raia fluviatilis.

Skates are very common in the Ganges, not only in the estuaries, but very far removed from the sea; for I have seen them at Kanpur, more than a thousand miles above the extent of the tide. In these upper parts of the river's course, and it is found also in the estuaries, the species most common has a strong resemblance to the *Raia lymma;* has nearly the same manners with the *Raia aquila,* and inflicts very dangerous wounds with the spine on its tail. It is, however, a very common fish, in the markets of Patna especially, and is not materially different in taste from a small skate or maiden ray, (*Raia batis.*) I always deferred taking a description, until I had an opportunity of having it drawn, and that opportunity never occurred. I cannot, therefore, give its specific character.

2d Species.—Raia sancur.

A *raia* with blunt teeth; with the body and tail unarmed, the latter provided with a fin; and with the head ending before in the segment of a circle.

The drawing of this fish was commenced, but never finished. It is common in the estuaries of the Ganges, both in fresh and salt water; but I have not seen it above where the tide reaches. It grows to nearly the same size with the skate, (*Raia batis,*) and for the table has nearly the same qualities. Except in wanting the prickle on the tail, it has a strong resemblance to the first species.

The *shape* of the whole animal, including the fins, approaches that of a rhomb, the pectoral fins forming an angle near the centre of the body. Independent of the fins, the body is oblong, and thicker, in proportion to its size, than that of the skate. The *colour* is an ugly dark brown above, and white beneath.

The *head* terminates in a blunt soft smooth snout. Near this are the *eyes,* which are large; and immediately behind each is a large triangular aperture, analogous, I presume, to our ear. The *mouth* is not large. The upper *lip* is a fleshy membrane arising from the fore part of the head, and ending behind in a semicircular sinus. Under the fore part of this membrane, at each side, is a large aperture in form of a crescent, probably a *nostril.* In both jaws there are various rows of blunt *teeth,* closely wedged together. The *apertures for breathing* consist of five slits placed behind each angle of the mouth, so as with it to form a kind of circle open towards the tail.

The *body* every where below, and towards the sides above, is quite smooth; but on the middle of the back is covered with small rough scales, The *vent* is behind the middle. The *tail* is more than twice the length of the whole body, and very slender. It is quite smooth, and tapers gradually from the root, which is flat, to the point, where it becomes like a thick thread. It is scabrous above; but is neither prickly nor armed with a spine.

There is no *dorsal fin,* but behind the head there is a light coloured hard protuberance, the rudiment probably of this organ. The *pectoral fins* edge the greater part of the body, and, forming a slight angle near the middle, are rounded behind. The *ventral fins* are much smaller than the pectorals, and edge the body between them and the tail. They are rounded behind, and their edges are indented. On the under side of the tail is placed a narrow *fin* without rays. It commences about three quarters the length of the body from the root, and runs along for about a space equal in length to the body.

II. Genus.—Squalus or Shark.

Fishes of the first order having on each side of the neck from five to seven breathing apertures, without gill-covers or membranes.

In the mouths of the Ganges sharks are exceedingly numerous, and occasionally, but rarely, come up as far as Calcutta. There would appear to be many species, usually divided by our seamen into two kinds,—one, with rounded bodies, called merely sharks; and the other, with flattened heads and chests, called ground-sharks. I have only had an opportunity of ascertaining two of the species, both belonging to the first kind.

1st Species.—Squalus carcharias?

A shark with one anal fin; with no aperture behind the eye in place of an ear; with triangular teeth, cut like a saw on the edge; and with the tail fin deficient near the point below.

This animal resembles very much the figure of the *S. carcharias*, given by La Cepede, (Tome I. Planche VIII. fig. 1;) but differs considerably from the figures given by Bloch (Ichth. Tome IV. Pl. CXIX.) and by Bonnaterre in the *Tableau Encyclopedique*, (Plate VII. fig. 20,) in both of which the tail fin entirely surrounds that member, in place of being deficient on the under part, towards its point. I must, however, observe, that the *Pala sorra* of Dr Russell (Indian Fishes, Vol. I. p. 9, fig. 14,) is a fish resembling this shark very strongly indeed, yet its caudal fin surrounds entirely the tail. We may therefore suspect, if this mark does not denote the difference of sex, that we have two species agreeing with the specific character usually given of the *S. carcharias*, or common shark. The specimen from which I took the description was very small, and differed from the full account given by La Cepede in the few following particulars, which can scarcely perhaps be considered as adequate to constitute a new species.

The *colour* of the fish is above green, and beneath whitish, with somewhat of a silver gloss.

The *nostrils* are situated nearer the eyes than the end of the head. On each side two parallel rows of *punctures* run from the nape to the first aperture for breathing. Before each eye there is another row, in form of a crescent, with the horns turned towards the sides of the head.

There is no longitudinal furrow on the *tail.*

The insertion of the *caudal fin* above is much shorter than below. The hinder *dorsal fin* is much smaller than the anal, and is shaped somewhat like a sickle.

2d Species.—Squalus pectinatus.

A *squalus* with a long bony snout, armed on each side with about 34 narrow teeth.

Pristis pectinatus, Transactions of the Linnæan Society, Vol. II. p. 278.

Saw-fish of the English.

This animal is said by the fishermen to grow to the length of from six to seven and a half feet, and to live entirely on weaker fishes. It would be needless to describe farther a fish already well known to the naturalist.

III. Genus.—Tetrodon.

Fishes of the first order having projecting jaw-bones, each divided as it were into two teeth, but there are no real teeth; and having no scales; but the body is slimy, and is capable, at the will of the animal, of being inflated.

The fishes of this genus are eaten by the poor, but are considered as indifferent food; nor does their quantity any where occasion them to be of any importance.

All the species described below have five *fins,* of which the *dorsal* is placed far back opposite to the *anal;* and the *pectorals,* which are small, are placed near the middle of the fish in respect to height.

The *rays* of all the fins, except the caudal, are so covered by a thick skin, and are so small, that they are numbered with difficulty and uncertainty, so that a great expectation of accuracy concerning the numbers stated should not be entertained. All these species have small horizontal *mouths* with fleshy *lips*, the *vent* behind the middle, and the *gill aperture* near the pectoral fins, and forming part of a circle.

1st Species.—Tetrodon fluviatilis. Plate XXX. Fig. 1.

A *tetrodon* with a body little compressed; with jaws of nearly equal length, but the superior by a little the longer; and with the back dusky-green divided, by yellow lines, into angular spaces like the plots in a garden.

This fish grows to about six inches in length, and is found in the fresh water rivers of the lower parts of Bengal. On this account it is called by the natives the river *Patoka*, to distinguish it from another found in ponds, and from another found in estuaries, for the term *Patoka* is generic.

This fish, as to *form*, when it inflates itself, has a moderately projecting belly, somewhat shaped like an egg, and tapering gradually to the tail. On the head, back, and belly, it is armed with small undivided prickles, which the animal can retract into some large pores in its skin. With respect to *colour*, the sides have large round spots, the lower parts are white. The fins are dusky, with two dark transverse bars across that of the tail. The eyes are yellow. When this fish has lived in dirty water, its belly becomes black, and its fins yellow.

The *head* slopes downward from the neck to the mouth, is shaped like an egg, is large and wider than the body, and ends bluntly. The *jaw*-bones do not protrude in opening.

The upper *lip* is rather the longest. The *tongue* adheres to the lower part of the mouth. The *palate* is covered with folds. The *nostrils* are far back, and separate. Each has a single aperture, with an erect leafy membrane, divided to the base into two blunt portions. The *eyes* are towards the upper and hinder part of the head, but far from each other. They are of a moderate size, and convex, with circular pupils. The *gill-covers* are overspread with flesh. I can only trace one ray in each of the *membranes* covering the gills, but this membrane is very thick.

The *thorax* is convex. The *abdomen* is capable of less distension than in most species of this genus. There is no *lateral line.*

The *back fin* is rounded, and contains sixteen rays, of which the first three are undivided; the second being shorter than the third, and longer than the first. The others are divided. The *pectoral* fins are rounded, and each contains seventeen divided rays. The *anal* fin is rounded, and contains fourteen rays. The tail fin is rounded, and contains eight divided rays.

2d Species.—TETRODON PATOCA. Plate XVIII. Fig. 2.

A *tetrodon* with the body scarcely compressed, but armed with simple thorns, and above blackish, with numerous yellow spots, and with the upper jaw the longest.

This fish is found chiefly in the estuaries, and is considered by the natives as the prototype of their genus Patoka, on which account they call it *the great Patoka,* and I have, therefore, applied their name as specific, especially as the term *great* would be improper, many other species being larger, for our fish seldom exceeds a foot in length: in the Ganges, however, we have no species so large. This fish has a considerable resemblance to the *T. sprengleri* and *T. testudineus* of Bloch; but, from the first, it differs in wanting the lateral appendants, by which that fish is distinguished; and from the second it is known by wanting longitudinal stripes on the sides, by the yellow spots on the back, and by a more prominent belly. The *Tetrodon kappa* of Dr Russell (Indian Fishes, Vol. I. No. 25) has also a considerable affinity to the Patoca, but is less prominent on the belly, and more so between the eyes, while the spots on the back are pale blue its belly yellow, and its back only rough, and devoid of visible prickles.

The *colour* of the *Patoca* above is blackish, with numerous large yellow spots on the back: below it is white. The fins are inclined to olive. The back and belly are *armed* with simple retractile prickles.

The *head* is unarmed and large, and slopes gently down to the mouth. The *jaws* do not protrude in opening; the under one is provided with an erect membranous veil. The *nostrils* and *eyes* are similar to those of the last mentioned species.

The *back* is arched, and its edge is rounded. The *belly*, when inflated, forms half a sphere. There is no *lateral line*.

The *back fin* is rounded, and contains eleven rays, of which the two first are undivided, and the first is the shortest. The *pectoral* fins are sharpish above. Each contains about sixteen rays. The anal fin is rounded, and contains ten rays, of which the first is the shortest, and, together with the second, is undivided. The tail fin is rounded, and contains ten rays, of which one on each side is undivided.

3d Species.—Tetrodon cutcutia. Plate XVIII. Fig. 3.

A *tetrodon*, with the body very little compressed, entirely unarmed, green above, and irregularly variegated with dusky and white below; with a large black circular spot on each side, where the tail commences, and with the under jaw the longest.

This fish seldom exceeds four inches in length, and is miserably tormented by idle children, who amuse themselves by inflating its belly. Every where watered by the Ganges and its branches, it is common enough in ponds and in small rivers.

The *cutcutia* is entirely smooth; and its belly, when inflated, forms more than half a sphere. The tip of the tail, fin, and the eyes, are red.

The *head* is rather flat and small, and ends bluntly. The *nostrils* are covered by undivided membranes. The *eyes* are large, and look downward. I can perceive no ray in the gill membranes.

The *back* slopes very gently from the middle of the fish in both directions, and is convex. There is no *lateral line.*

The *back fin* is rounded, and contains twelve rays. The *pectoral* fins are rounded, and each contains about twenty-one rays. The *anal* fin is rounded, and contains eleven rays. The *tail* fin is rounded, and contains eight rays, which project a little beyond the membrane.

4th Species.—TETRODON CARIA.

A *tetrodon* with the body very slightly compressed, entirely unarmed, above green, with irregular dusky marks, and below black; and with jaws nearly equal in length.

This fish is found in the river Kosi, and differs so little from the *cutcutia*, that I cannot take upon myself to say, whether it is a distinct species or a mere variety. Besides the circumstances related in the specific character, it wants the large spot on the sides, by which the cutcutia is distinguished, and its sides are reticulated with narrow black lines. The head and back slope more. The nostrils are tubular, and capable of being retracted. The rays of the dorsal fin are about eleven, of each pectoral about eighteen, of the anal about ten, and of the caudal eight.

5th Species.—Tetrodon gularis.

A *tetrodon* with the body very slightly compressed and entirely unarmed, above green, irregularly varied with dusky, and beneath black, with a large black spot on each side, with a red mark on the throat, and with jaws nearly equal in length.

This fish is found along with the preceding species, and differs only from it in the particulars mentioned in the specific character. Its being, therefore, a mere variety, is still more probable. I have been under the necessity of translating the specific name given by the natives, *Kanthawaleh Phoksha*, being altogether inconvertible to any word of decent length.

6th Species.—Tetrodon tepa.

A *tetrodon* with a compressed prickly body, and with a tail fin of an equal crescent shape.

This fish is found in the estuaries, where it grows to be about a foot in length, and is a very distinct species from any above mentioned; but it has a very considerable affinity to the *T. sceleratus* of La Cepede, (*Hist. des Poissons*, Tome I. p. 476,) which is said to be poisonous; in fact, the natives of Bengal reject the *Tepa* in food. The accounts of the *T. sceleratus*, however, are still very imperfect. The name *Tepa*, (*Tengpa*,) it must be observed, is generic, but all the other fish to which it is applied have some other distinguishing appellation. I have therefore confined it to this species, as the prototype of the native genus *Tepa*.

This fish in *form* is conical, compressed, and on the back and belly has a space surrounded by a margin, that contains undivided prickles capable of being retracted. When inflated, the belly forms a considerable segment of a sphere. The *colour* is olive above, and white below, with a golden gloss on the sides.

The *head* is very large, and very blunt at the nose. The *jaws* are sharp, and the under one is the shortest. Each *nostril* has two small openings placed in a circular depression, without any membranous appendant. The *eyes* are large, stand high up, and are of an oblong form.

The *back* slopes but little. The *lateral line* begins at the nostril, passes under the eye, ascends to the neck, and there forms an elevated ridge, which bounds the prickly portion of the back. At the dorsal fin it bends down, and runs through the middle of the *tail*, which is quadrangular.

The *dorsal fin* is falcated, and contains fourteen rays. The *pectoral* fins are small; and each contains fourteen rays, of which the extreme ones are undivided. The *anal* fin is falcated, and contains eleven rays, of which the two first are undivided, and the last is divided to the root. The *tail* fin consists of two sharp lobes, of which the upper is the longest, and contains thirteen rays, of which three on each side are undivided.

IV. Genus.—Syngnathus.

Fishes of the first order with a very small toothless mouth, situated at the extremity of a long cylindrical tube, composed of the jaws united.

All the fishes of this genus, which I have seen in the Ganges, are from seven inches to a foot in length, and not thicker than a strong goose's quill. Being rather uncommon, and of an uninviting appearance, they are seldom eaten, and in an economical view are of no importance. Their *heads* have no scales, and the *tubes*, in which they terminate, are slightly compressed, and nearly of the same length with the head. Their *bodies* are as it were jointed, each joint being encircled by a hard scale, which covers the whole. The number of *angles* in the body varies in the different species; but, in the tail, which tapers to a point, all have four acute angles. All the species which we have belong to the second division of La Cepede, having four fins, one on the back, two pectorals, and that on the tail.

1st Species.—SYNGNATHUS CUNCALUS.

A *syngnathus* with four fins; and with seven angles on the body, which contains eighteen joints, while the tail has twenty-five.

I found the *cuncalus* in the estuaries near Calcutta. It has a strong affinity to the *S. pelagicus* of La Cepede, (*Hist. des Poissons*, Tome II. p. 40,) but differs in the number of joints and of the rays in the fins. In the month of September the female fish has her belly, from the throat to the vent, covered with eggs adhering fast to the scales.

The fish in *colour* is green above, and white beneath, with a gloss of silver on the sides. On the back several dusky lines, intersecting each other, form a net-work. On each side is a longitudinal stripe formed of numerous dots. The eyes are silvery.

The *head* is very small, and narrower than the body, compressed at the sides, and straight above. The *nostrils* are far back, and each contains one small aperture. The *eyes* are far up, globular and protuberant, with circular pupils. The *mem-*

brane of the *gill-covers* has two rays, and closes entirely the *aperture*, adhering firmly by its whole edge.

The *back* is straight and concave, edged on each side by a ridge, which is notched at each joint. The *sides* are edged in the same manner with the back, each forming an obtuse angle in the middle. The belly is straight, and like the sides has an obtuse angle in the middle. The four acute edges run along the *tail*, the three obtuse ones terminate at the *vent*, which is near the middle. On each side are two *lateral lines*, both straight, and one placed above, while the other is below the obtuse angle of the side. The former terminates at the end of the dorsal fin, the latter before the vent.

The *back* fin occupies three joints of the back and seven of the tail; and contains fifty undivided rays. The *pectoral* fins are very small, rounded, and near the middle of the side in respect to height. Each contains sixteen rays. The *tail* fin is rounded, and contains nine rays.

2d Species.—Syngnathus carce.

A *syngnathus* with four fins, and seven angles on the body, which contains thirty-two joints, while the tail has eighteen.

This species I have observed only in the tide-ways. The natives call it the Crocodile's Tooth-pick, or Tooth-cleaner.

The *colour* is greenish-brown, with some obscure pale transverse belts on the back, and ten or eleven spots on each side at the lower edge of the scales, by which the joints are surrounded.

The *head* is compressed, nearly straight, narrower than the body, small, and armed on the throat with a short blunt prickle. Each *nostril* contains two minute apertures, placed near the *eyes*, which are small, protuberant, and high on the head. The pupils are circular. The rays of the *gill-covers* are very indistinct, and the *openings* of the gill-covers are very small, and placed on the sides of the neck.

The *back* is straight and concave, bounded on each side by an acute, very entire, angle or ridge. The *sides* have a blunt longitudinal ridge along their middle. The *belly* is straight, and an angle or ridge runs along its middle to the vent, where, dividing into two, its branches unite with the sharp ridges, separating the sides from the belly. On the fore part of each side there are two *lateral lines*, dotted with white. The *scales*, or *belts*, by which the animal is surrounded, are striated and sprinkled with dots.

The *back-fin*, placed near the middle of the fish, contains twenty-five rays. The *pectorals* and *tail fin* are rounded, and contain each ten rays.

3d Species.—SYNGNATHUS DEOCATA.

A *syngnathus* with a six-sided body, and four fins.

The *S. barbarus* is said, in their descriptions by Bonnaterre, (*Tabl. Encycl.* p. 31, T. XXI. f. 74,) and by La Cepede, (*Hist. des Poissons,* Tome II. p. 48,) to have no tail fin ; but, in the figure given by the former, it is represented with that organ, and has a strong resemblance to our fish.

The *deocata* I have found in the Tista, Kuwarlayi, and other rivers of Puraniya, or Mithila, in the north of Bengal and Behar. It is of a brown *colour*, with the sides of the belly below the lateral lines beautifully variegated with red and blue; but, when the female breeds, these bright colours disappear, except on the margins of the belly, which become yellow, and are dilated, in order to receive two rows of globular yellow eggs, each row containing about twenty, which adhere to the belly, until hatched.

The *head* is not compressed. The *gill-membrane* adheres all round, and entirely closes the *aperture.*

The *back* is straight, and surrounded by a sharp ridge. An obtuse angle runs along each *side.* The *belly* is flat, and surrounded by a sharp ridge. The *lateral lines* are indistinct, but are in the middle of the upper face of the sides. The *vent* is near the middle. Before it are sixteen *joints,* and behind it thirty-two.

The *dorsal* fin occupies two joints of the back and four of the tail, and contains about thirty rays. The *pectoral* fins are rounded, and each contains about fifteen rays. The *tail* fin is oval, ends in a sharp point, and contains about seven rays; but their number cannot be distinctly traced.

ORDER II.

FISHES HAVING THE DORSAL SPINE OF BONE, AND WANTING VENTRAL FINS.

I. Genus.—UNIBRANCHAPERTURA.

FISHES of the second order without pectoral fins; with a body shaped like a serpent, and having only one opening to the gills situated under the throat.

I have retained the name given to this genus by La Cepede, although it is both long and barbarous, and need not have superseded the *Synbranchus* given by Bloch, which, although far from harmonious, is at least shorter. The name of La Cepede is, however, very significant, and expresses the chief mark by which the genus is distinguished.

1st Species.—UNIBRANCHAPERTURA CUCHIA. Plate XVI. Fig. 4.

An *Unibranchapertura* destitute of fins, and spotted with black.

The *Dondoo Paum* of Russell (Indian Fishes, Vol. I. No. 35) has a strong affinity to our fish; but has not the black spots, and may perhaps be the *Unibranchapertura lævis* of La Cepede, (*Hist. des Poissons*, Tome V. p. 657.)

This fish I found in the rivers and ponds of the south-east parts of Bengal. It is said to grow to two feet in length, and six inches in circumference: but those that I have seen were shorter and thinner in proportion to their length. Europeans eat the *cuchia* as an eel; but the natives reject it, and imagine that its bite is fatal to cattle, although less powerful on the human kind,—a supposition highly improbable.

The whole *form* of the animal, having no vestige of a fin, resembles strongly a serpent. This fish is cylindrical, devoid of scales, and lubricated with slime. The *colour* above is dark green, below a dirty pale red. On every part are scattered small round black spots, and short yellowish lines. Two parallel pale lines run forward from the shoulders, and at the eyes join, at an acute angle, two others coming from the sides of the chest. Parallel to the last are two others, one on each side beneath the lateral line.

The *head* is blunt and of a moderate size, but swells out when the animal is irritated. The *snout* is depressed, short, and retuse, projecting a little beyond the under jaw. The *nostrils* are near the end of the snout, very small, and each has only one aperture. The *eyes* are near the upper part of the head, and small, as is also the *aperture* to the gills.

The *lateral line* forms an arch before, and then runs straight along the middle of the side. The *tail* ends in a sharp point, is compressed, and has sharp edges above and below.

II. Genus.—Murænophis.

Fishes of the second order, with a body like a serpent, but without scales, and devoid of pectoral fins, with the fins of the back vent and tail united into one, and with an opening to the gills on each side of the neck.

The genus, of which I am now treating, by Bloch has been called *Gymnothorax*, which La Cepede was scarcely justifiable in changing to *Murænophis;* but, as the *Muræna* of the ancients belongs to this genus, the name might have been appropriated to it, while the name *Anguilla* might have been bestowed on the genus, which La Cepede calls *Muræna*, as the common and well known eel is no doubt the *Anguilla* of the Latins.

1st Species.—Murænophis sathete.

A *murænophis* with the dorsal fin commencing at the neck, with sharp teeth placed at a distance from each other in both jaws, which are nearly of equal lengths, and with an unspotted body.

The *sathete* I have found in the estuaries near Calcutta. It is a very ugly eel, which grows to five or six feet in length, and is of a brownish black *colour*, with the under parts of the head of a dirty yellow.

The *head* is oval, and small. The nostrils are double, the foremost being tubular. The *eyes* are small, and far forward. The *teeth* are strong. The *gill-membrane* is thick, large, and striated with small rays, the number of which cannot be properly ascertained. The *openings* to the gills are small.

The *body* is very slender in proportion to the length, nearly round, and very slippery. The *lateral line* runs down the middle of the side. The only *fin* runs from the neck round the tail to the vent, and has such a thick skin that the number of rays cannot be ascertained.

2d Species.—MURÆNOPHIS TILE.

A *murænophis* with the dorsal fin commencing at the nape; with a row of remote sharp teeth in each jaw, and one large tooth in the palate; with the body divided by lines into small squares, and variegated with irregular yellow blotches and black dots; and with the under jaw shortest.

This is an ugly eel, seldom growing to above eighteen inches in length. It is found in the estuaries near Calcutta, and has the strongest affinity to the *Gymnothorax reticulatus* of Bloch, (*Ichth.* Tome XII. p. 71, f. 416,) or *Murænophis reticularis* of La Cepede, (*Hist. des Poissons*, Tome V. p. 628,) who state, however, that the palate is smooth, (*uni* Bloch, *lisse* La Cepede,) which I presume could not be said of the Gangetic fish, on account of the remarkable fang which it has, and by which it acquires a strong analogy with poisonous serpents. Owing to this circumstance, it has a great affinity with the *Murænophis grisea* of La Cepede, (*Hist. des Poissons*, Tome V. p. 629,) and with the *Calamaia paum* of Russell, Indian Fishes, Vol. II. No. 32.) Indeed, it agrees in most particulars with the description given by Dr Russell, but the figure differs in many points.

The *head* is small, oval, and blunt. The *nostrils* are double; the foremost are tubular, and placed near the jaw; the hinder are near the eyes. The *teeth* are strong, and bent backward. The *mouth* is large, reaching behind the *eyes*, which are small, round, and yellow. There are no *gill-covers*; but the *mem-*

brane is large, and very much dilatable. It is striated with several rays; but their number cannot be exactly ascertained.

The *body* is not compressed, but nearly round. It is slippery and naked, although numerous depressed lines, crossing each other, form it into little squares like scales. The *lateral line* runs straight down the middle of the side. The *tail* tapers to a point, is compressed, and has a sharp edge above and below.

The *fin* is broad, extends from the gill openings round the tail to the vent, and is so thick that the number of slender rays which it contains cannot be ascertained.

III. Genus.—Ophisurus.

Fishes of the second order, with indistinct pectoral, dorsal, and vent fins, but having no fin on the tail, and those on the back and behind the vent are very long and narrow, with a body formed like a serpent, and with an opening to the gills on each side of the neck.

The three species of this genus, which I have found in the Ganges, for eels are rather beautiful animals, and are eaten by many of the natives, most of whom reject the others, although all are used very indiscriminately by Europeans.

These three eels in *form* are very slender, and nearly cylindrical in the body, but taper to a point at both tail and *head*, this member terminating in a sharp snout, which projects considerably beyond the under jaw. Their *nostrils* are solitary tubes near the end of the snout. The *eyes* are small and far forward. The *teeth* in both jaws are small. The *openings* of the gills are small and close to the pectoral fins. No rays can be distinctly traced in the *membrane* covering the gills, which is thick. The *gill-covers* are flexible, and covered with skin. There are no *scales*, but the body is not so slimy as that of the common eel. The *vent* is placed before the middle.

1st Species.—OPHISURUS HIJALA. Plate V. Fig. 5.

An *ophisurus* with the upper jaw the longest, and with a row of pale spots on the fore part of the lateral line.

This I found in the rivers of Bengal, both salt and fresh. The natives have an imagination that it proceeds from the ear of the porpoise, and one of its names implies this circumstance. It does not grow to above eighteen inches in length.

The *colour* of the under parts is white; the upper are thickly covered with green dots, which, before the vent, descend lower than the lateral line, but towards the tail this line separates the dotted from the white part. The pale spots on the lateral line reach as far back as the vent, and are round and contiguous.

The *head* is as thick as the body. The *eyes* are high.

The *lateral line* is straight, and higher a little towards the shoulders than on the tail.

The *dorsal* fin extends from the head to within a tenth of an inch of the point of the tail, contains from two hundred and thirty to two hundred and forty undivided rays, nearly of an equal length, and can be lodged in a longitudinal furrow. The *pectoral* fins are very small and rounded, and are near the middle of the sides: each contains eight divided rays. The *anal* fin runs from the vent almost as far back as the dorsal, and contains about one hundred and seventy fine undivided rays, nearly of an equal length.

2d Species—OPHISURUS BORO.

An *ophisurus* devoid of spots or lateral line, with the upper jaw the longest, and with about three hundred and thirty rays in the back fin.

This fish I have found in the estuaries near Calcutta, where it grows to about eighteen inches in length.

The *colour* is above black, with green dots; beneath very pale yellow. The *eyes* are silvered.

The *head* is wider than the body. The *eyes* are near the middle of the head in respect to height.

The *pectoral* fins are minute and rounded, and each contains about eleven rays. The *anal* fin has about two hundred and seventy.

3d Species.—Ophisurus harancha.

An *ophisurus* with upper jaw the longest; without spots, but having a lateral line; and with about three hundred and forty-five rays in the back fin.

This fish I found also in the estuaries near Calcutta, and it differs little from the *Boro.*

The *lateral line* runs straight along the middle of the fish. The *vent* fin can be hid in a longitudinal furrow, and contains about two hundred and forty rays. The *pectoral* fins end in a sharp point, and each contains about twelve rays.

IV. Genus.—Muræna.

Fishes of the second order with a serpent-like form; without scales, but covered with viscid slime; and having fins on the tail, sides, and back, and behind the vent, the two latter being unarmed.

I have already mentioned the impropriety of bestowing the name Muræna on a genus, to which the Muræna of the Latins does not belong: but I leave the correction to some publisher of a system.

In Bengal fishes of the eel kind are seldom eaten except by the lower classes of natives, and by Europeans; nor, in an economical point of view, are they of any considerable importance; as, although of uncommon variety, and found in every part of the Gangetic provinces, they are no where in such abundance as to be of consequence. Perhaps, however, the smallness of the number taken may be owing to the want of demand, and the consequent neglect of the fishermen. I have not been able to trace the source of what Pliny states (*Hist. Mundi*, L. IX. c. iii.) concerning the immense length of the eels in the Ganges, which he says are three hundred feet long. All that I have seen or heard of were of dimensions, that, in Europe, would be considered moderate; nor do I know any other animâl in thé Ganges of dimensions that could have given rise to such an exaggeration.

1st Species.—Murnæa anguilla.

A *murœna* with the under jaw the longest; with about a hundred rays in the anal fin, which is united with those of the tail and back; and without spots on the back or tail.

Murœna anguilla, La Cepede, *Hist. des Poissons*, Tome II. p. 225.

Chowloo Pamoo or *Malgumaroo*, Russell's Indian Fishes, Vol. I. No. 31.

This eel, which is exactly the same with the common eel of Europe, I found in the marshes communicating with the Tista. It is astonishing that La Cepede should make such a fuss about an animal which has every appearance of a serpent without the vivid colours, by which most of these reptiles are adorned. It is an irritable creature, swelling its head whenever angered; and constantly, when it can, buries itself in putrescent carcasses. In our Indian marshes its upper parts are of a dirty green *colour*, variegated with paler and darker shades: beneath it is of a dirty yellow.

2d Species.—Muræna maculata of La Cepede?

A *muræna* with the under jaw the longest; with a green body variegated with black irregular marks; and with the back, tail, and anal fins, united into one.

La Cepede (*Hist. des Poissons*, Tome II. p. 265) can scarcely be excused for changing the name *guttata* given by Forskael to *maculata*. The eel, which these authors describe, is found in the Nile, and I found this in the Ganges, where that river passes through Behar. The description given by La Cepede is so imperfect, that I am uncertain whether the animal of the Ganges belongs to the same species with that of the Nile.

The Gangetic fish is a very ugly animal, and, when old, is said to grow very thick in proportion to its length; that is, when full grown, it is about two and a half cubits in length, and one cubit in circumference; but in this, I suspect, there is some mistake, as the largest which I procured, and which was three feet two inches long, was only six inches and a half round.

The *colour* of the lower parts is white.

The *head* is wider than the body. The under *jaw* is a little longer than the upper. The *nostrils* are double: the foremost are close to the jaw, being tubular. The hinder apertures are near the *eyes*, which are high.

There is no *lateral line.*

The *back*, *vent*, and *tail* fins are so thick, that the number of rays cannot be ascertained. Each *pectoral* fin contains twenty-four.

3d Species.—Muræna vamos.

A *muræna* with the upper jaw much longer than the under, with the back, tail, and anal fins united, and without any remarkable spots, or a lateral line.

This eel is found in the estuaries of the Ganges, and is said to grow to seven and a half cubits in length, although none that I saw exceeded half that dimension. Above it is of a dirty brownish green; below the *colour* is dingy white.

The *head* is oval, and broad in proportion to the body, being nearly twice as wide. The upper *jaw* is blunt, and projects far beyond the under, which is sharp-pointed. It has tubular *nostrils.* The *eyes* are small, and high up.

The *body*, in proportion to its length, seems slenderer than that of the common eel. The *lateral lines* can scarcely be traced.

The number of rays in the *fins*, owing to their slenderness, and to the thickness of the skin, cannot be ascertained. The *pectoral* fins are rounded, and very short.

4th Species.—Muræna bagio.

A *muræna* with the upper jaw by much the longest; with the back, tail, and anal fins united into one, and with the body of a dirty brown above, having no remarkable spots, but with a distinct lateral line.

This eel has a strong resemblance to the *Taleo paum* of Dr Russell, (Indian Fishes, Vol. I. No. 36,) in the disposition of the teeth, the mark by which eels can be best arranged, although I was not aware of the circumstance in time to avail myself of it in describing most of the species. Dr Russell's fish is of a shining golden colour.

The *bagio* is found in the estuaries of the Ganges, and

grows to eighteen inches or two feet in length. The under parts are of a white *colour*, and, where this unites with the brown of the back, the sides are covered with dots.

The *head* is narrow and long, and ends in a sharp point. The *mouth* is very wide, and opens straight back. The *jaws* do not protrude in opening, and are blunt and covered by *lips*. In the fore part of each jaw are about six long sharp *teeth*. In the sides of the upper jaw there are none; but in those of the under jaw there are many small ones. In the middle of the palate there are three rows. The teeth in the side rows are small and irregularly crowded; in the middle row they are placed at a distance from each other, are large, compressed, and divided into lobes. The *eyes* are high, large, and oblong, with circular pupils. The *nostrils* are double, the anterior apertures being tubular. The *gill-covers* are small. The *membrane* which ties them down is long and thick, and streaked with several fine rays, which cannot be distinctly numbered, and which leave a small crescent-shaped *opening* below each pectoral fin.

The *body* is slightly compressed. The *lateral line* runs straight and high on the side, but highest before. The *vent* is before the middle.

The *fin*, composed of the dorsal, anal, and tail fins, extends from the nape to the neck, and has about two hundred and sixty rays above, and two hundred and twenty below. The *pectoral* fins are low, small, and sharp-pointed, each containing twelve rays.

5th Species.—Muræna raitaborua.

A *muræna* with the upper jaw longest; with the colour above purple with black dots; and with the back, tail, and anal fins distinct.

On account of the number of fins this eel is nearly allied to the fish described by Russell, (Indian Fishes, Vol. I. No. 34,) which wants the dots on the body, and has three apertures on each side of the nose.

This eel seldom exceeds a foot in length, and is less disgusting in appearance than most fishes of the kind. The under parts are white.

The *head* is small, oval, and sharpish. The *mouth* is small. In each *nostril* there are two apertures; neither of them tubular. There are strong *teeth* in both jaws. The *eyes* are small, high, and far forward. There are no *gill-covers;* but a very dilatable membrane, which is striated with some slender rays, that cannot be exactly numbered, and leave a small round *opening* at each pectoral fin.

The *body* is not compressed. The *lateral line* runs straight along the middle of the side. The *vent* is behind the middle. The *tail* tapers to a blunt end.

The *back fin* is behind the middle, and contains thirty-two rays of nearly equal length. The *pectoral* fins are minute and rounded, and each contains eight rays. The *anal* fin commences near the vent, and contains fifty rays of nearly equal length. The *tail* fin is rounded, and contains sixty rays.

V. Genus.—Macrognathus.

Fishes of the second order, covered with very minute scales, somewhat like a serpent, armed with prickles before the back and anal fins, and having a fin on the tail.

The fishes of this genus were formerly included among the *Ophidiums;* and, as the species of both were not very numerous, one genus might, perhaps, have included the whole without inconvenience. While there was only one genus, the specific names *aculeatum* and *armatum*, given to two species, were very proper; because several other species were devoid of prickles; but since the division, as all the species of Macrognathus have numerous prickles, these names have become absurd, and should be changed by the first systematic writer.

The fishes of the *Macrognathus* genus have less of a disgusting appearance than those called *Murœna,* and are more sought after by the natives; the highest of whom in Bengal make no scruple of eating them; and by Europeans they are esteemed the best of the eel kind.

All the species of *Macrognathus* are long slender fishes, considerably compressed at the sides. They are slippery, with very minute smooth scales, and at both ends terminate in a point.

The *head* is narrow, compressed, smooth, and lengthened out into a snout, which tapers to a sharp end. Their *mouths* are small, and situated under the snout, at some distance from the extremity. Two of the species have small *teeth* in both jaws; but in the third these organs are scarcely, if at all, observable. The *eyes* are small. There are always two tubular *nostrils* near the extremity of the snout; and in some species there are other two apertures towards the eyes. The *gill-covers* are large and distinctly seen, and do not conceal their *membranes,* which cover the sides of the throat. The *openings* to the gills are large, and one is placed on each side of the neck. The edges of the *back* and *belly* are convex; those of the *tail* are sharp above and below. The *lateral lines* are placed high, and are straight and smooth.

The *back* and *anal* fins are linear; and the former is placed far back along tail. Before each are several *prickles,* capable of being erected, or of being laid back into a furrow, by means in general of a small membrane. Their number varies in different individuals of the same species, but is always by far the greatest on the back. The *pectoral* fins are placed low, and are very short and rounded; as is also the case with the *tail* fin. The number of rays in the fins cannot, in general, be ascertained with exactness, owing to their slenderness, and to the thickness of the skin.

That part of the generic character given by La Cepede, (*Hist. des Poissons*, Tome II. p. 283,) which states the tail fin to be distinct from those of the back and vent, is objectionable, as in one of the two species, which he describes, the three fins are united.

1st Species.—MACROGNATHUS ARMATUS. Plate XXXVII. Fig. 6.

A *macrognathus* with fins of the tail, vent, and back united; *Macrognathus armatus*, La Cepede, *Hist. des Poissons*, Vol. II. p. 286.

Ophidium simack, Walbaumii, *editio Artedii.*

This species, owing to the length of the body, and to its being less compressed than the others, approaches nearest the *Murænas.* It is often found two feet in length, and is esteemed the best for eating. It is found in the rivers of Bengal, both salt and fresh.

The *colour* of this fish is green, variegated with many black spots and dots. In such as that from which the drawing was taken, being young, a winding longitudinal stripe runs back from the snout along the lateral line, uniting the spots into irregular transverse bands. The breast and back fins are spotted with black.

The *head* is half oval, and flat above. The *mouth* is larger, and the *snout* shorter than in the other species. The *lips* are thick, with scarcely any bone; the inferior one is very blunt. In both *jaws*, which do not protrude in opening, are many sharp *teeth*, nearly equal in size, and closely imbricated. The *tongue* is blunt, and free from the under jaw. The *palate* is smooth. The *nostrils* are single. The *eyes* are far back in the middle of the head. The gill-*covers* are scaly.

The *tail* is blunt. The *vent* is near the middle. The *prickles* before the back fin are about thirty-seven; before the anal fin they are two in number.

The *back, tail,* and *anal* fins can only be distinguished by the rays of the tail being a little longer than the others. Those in the back part are from eighty-three to eighty-five, those in the tail seventeen, and those in the vent part from eighty-eight to ninety. Each *pectoral* fin contains twenty-one divided rays.

2d Species.—Macrognathus aculeatus.

A *macrognathus* with several eye-like spots on the back fin, which is distinct from that of the tail.

Macrognathus aculeatus, La Cepede, *Hist. des Poissons*, Tome II. p. 284.

Ophidium aculeatum, Bloch, *Ichth.* Tome IV. p. 60, Planche CLIX. fig. 2; Bonnaterre, *Tabl. Encycl.* page 41, Pl. XXVI. fig. 90.

This fish, if the native names are rejected, should be called *M. ocellatus.* Every where in the Gangetic provinces it is found in ponds, and grows from six to twelve inches in length. The *colour* of the upper parts is greenish with black dots, of the under yellowish. Above the lateral line, on each side, is a broad whitish stripe. At the roots of the dorsal fin are from three to seven eye-like spots, consisting of a black centre surrounded by a yellow ring. On the back and anal fins are scattered black dots, and two yellowish longitudinal stripes pass on the former, one on the margin, the other on the middle. The tail fin is spotted with yellow.

In neither jaw are there any *teeth.* The *lips* are fleshy, with very minute bones. Each *nostril* has two apertures. The *eyes* are situated high. The gill-*covers* are scaly. The *membrane* contains about five rays.

The *vent* is behind the middle. Before the back fin are about twenty distinct *prickles;* before the anal fin there are three.

The *back* fin is arched, and contains about fifty-one rays. The *pectoral* fins are low, and each contains twenty rays. The *anal* fin is arched, and contains about forty-eight rays. The *tail* fin is distinct from both back and anal fins, and contains fourteen rays.

3d Species.—MACROGNATHUS PANCALUS. Plate XVII. Fig. 7.

A *macrognathus* with the tail fin distinct; and with many black and white dots on the body.

This is a beautiful fish, which is found in tanks, and grows to from four to six inches in length. The *colour* above is green, and below white, marked on the sides with many black dots and small white spots. The hinder part of the back fin, and those of the vent and tail, are brownish with many black dots, forming transverse stripes.

The jaws contain minute *teeth.* Each *nostril* has only one aperture. The *scales* on the gill-covers are not visible. Each of the *membranes* covering the gills contains four rays.

The *tail* is sharp. The *scales* are very minute. The *prickles* before the back fin are about twenty-five; before the anal fin they are three in number.

The *back* fin is arched, and contains about forty rays. Each *pectoral* fin contains twelve; and the *anal* fin, which is arched, has about twenty-five.

VI.—Genus Trichiurus.

Fishes of the second order, with bodies very much compressed, and ending in a tail, which is drawn out to a long narrow point, and has no fin.

1st Species.—Trichiurus lepturus.

A *trichiurus* with the under jaw longer than the upper.

This fish, by the English in India, is called *Bombelo,* has a rich taste, is easily dried in the sun, and is found in great quantities in our estuaries. Bloch, (*Ichth.* Planche CLVIII.) Bonnaterre, (*Tabl. Encycl.*) and La Cepede, (*Hist. des Poissons,* Tome II. p. 182, Pl. VII. fig. 1,) allow only one hundred and seventy rays in the dorsal fin of their *Trichiurus lepturus,* which frequents the fresh waters of South America; and La Cepede's figure represents the lower jaw as terminated by an acute turned-up point. This leads me to suspect that their fish and ours are different species, although no doubt very similar; and I suspect the same of Dr Russell's *Trichiurus lepturus,* (Indian Fishes, No. 41,) which has one hundred and twenty rays in the back fin, and a sharp under jaw. Perhaps, therefore, three fishes have been included under one name; but this being uncertain, I think it unnecessary to multiply distinctions, and shall only add a full description, so that those who have an opportunity may judge how far any differences to be observed in different places may be adequate to distinguish these kindred fishes into different species.

Our Bengal fish is about a foot in *length,* very long, and much compressed, so as to resemble the blade of a very sharp-pointed sword. It is every where of a bright silver *colour.*

The *head* is very much compressed, and devoid of scales. The under *jaw* is the longest, and ends very blunt. In both jaws are long distant *teeth*, and at the end of each are two longer than the others. The *eyes* are high. Each of the *nostrils*, which are near the eyes, has one aperture. The *gill-covers* are large, and are finely cut on the edges. The *openings* of the gills are large.

The *body* is without *scales*. The *back* is arched, and sharp-edged. The *belly* is straight, and blunt-edged. The *tail* has both edges sharp, beneath is indented with about seventy-four small prickles, and terminates in a very long slender point, which, towards its end, resembles a bristle. The *vent* is before the middle. The *lateral line* on each side is bent down from the nape, and is below the middle.

The *back* fin is very long, and contains one hundred and fourteen rays. The *pectoral* fins are sharp above, and very short, and each has eleven rays. There are no other fins.

VII.—Genus cynoglossus.

Fishes of the second order, with both eyes on one side of the head, and with a flat body, formed for swimming on the side opposite to the eyes.

1st Species.—Cynoglossus lingua.

A *cynoglossus* with the eyes on the left, with no pectoral fins, and those of the back and tail united with that behind the vent.

This fish has the utmost affinity to my *Achirus cynoglossus*, which will be afterwards described, to the *Achirus bilineatus* of La Cepede, and to the *Jeree poloo* of Russell; but, as it totally wants the ventral fins, it must be placed in this order.

By the English of Calcutta it is called a *Sole*, and equals in quality and size this most valuable fish; but in that city it is not very common, as it is found only in the estuaries that are strongly impregnated with salt. The specific name is a Latin translation of that given to it by the natives. The generic appellation is taken from its *shape*, which is long, oval, very much flattened. Above it is rough, and below smooth. The *colour* of the upper or left side is brown with some gloss of gold, and is clouded with irregular black spots. The right side is white.

The *head*, viewed from above, is oval, small, and blunt. The *mouth* is small, at some distance behind the extremity of the nose, and passes very obliquely towards the eyes. The *lips* are fleshy: the upper one is the longest, incurved, sharp, and without bones; the under one is blunt. There are no *teeth*. One *nostril* is placed between the eyes, and the other between the lower eye and mouth. The *eyes* are near each other and small.

The *tail* is sharp. On the right there is no *lateral line*, on the left there are two; one near the back, the other running straight along the middle. The *vent* is close to the throat, a little obliquely towards the right side.

There is only one *fin*, which extends from the crown of the head round the tail to a little before the vent, where it takes a slight bend to the left. A small elongation of rays marks in some degree the part belonging to the *tail*, which contains eight rays, while the part on the *back* contains about one hundred and forty, and that toward the *vent* has one hundred and eleven.

ORDER III.

FISHES HAVING THE DORSAL SPINE OF BONE AND THE VENTRAL FINS PLACED BEFORE THE PECTORALS.

I. Genus.—BATRACHOIDES.

FISHES of the third order, with a large depressed head, and with a very large mouth, which is surrounded by tendrils.

1st Species.—BATRACHOIDES GANGENE. Plate XIV. Fig. 8.

A *batrachoides* with a body devoid of scales, but scattered over with short filaments; with three prickles on the back, and four on each side of the head.

This fish has a strong resemblance to the *Batrachoides tau* of La Cepede, (*Hist. des Poissons*, Tome II. p. 452;) which, by Bloch, (*Ichth.* Tome II. p. 150, Pl. LXVII. fig. 2 and 3) has been classed in the genus *Gadus* along with the cod, haddock, whiting, &c. from which it has been most properly removed. I doubt much whether our fish be different from the *Cottus gruniens*, although the number of rays in its fins differs somewhat from that stated by Bonnaterre, (*Tabl. Encyclop.*) and although a *Cottus* should have the ventral fins placed behind the pectorals; but in Bloch's figure of the *Cottus gruniens* the fins are placed as in the fish I am now describing.

The *gangene* is found in the salt-water estuaries of the Ganges, is an ugly animal, does not exceed a span in *length*, and when frightened emits a remarkable creaking noise. It is of a conical *shape*, slippery, and its *colour* is variegated with black and brown as it were in clouds. The fins are yellowish-brown variegated with black. The eyes are reddish.

The *head* is depressed, blunt, very large, and has scattered on it many soft short filaments, of which those nearest the jaws are longer and branched. At the upper angle of the gill-covers, on each side, are four recumbent *prickles.* The gap of the *mouth*, situated at the extremity of the head, is very wide. The *jaws* are depressed, do not protrude in opening, and the under one is the longest. The lips are *fleshy.* In each jaw there is a belt of numerous *teeth* set close without order. The *tongue* is smooth. The *eyes* are high and far forward; they are remote from each other, small and circular. The *nostrils* are not distinguishable. The *gill-covers* are small and covered with skin. Their *membranes* are very long, and each contains four rays. The *apertures* are of moderate size, and close to the pectoral fins.

There is no *lateral line* nor *scales*, but over the body are scattered many short soft filaments. The *vent* is near the middle.

There are two fins on the *back:* the first is small, and contains three prickles; the hinder one is long, and contains nineteen soft rays. The *pectoral* fins are rounded, and inserted into a projecting fleshy process: each contains twenty-two soft rays, every one of which is twice divided into two. The *ventral* fins are situated on the throat, are distinct from each other, and commence under the membranes of the gill-covers: each contains one strong prickle and five soft rays. The *vent* fin has sixteen rays. The *tail* fin has about fourteen, and is rounded.

ORDER IV.

FISHES HAVING THE DORSAL SPINE OF BONE, AND VENTRAL FINS PLACED IMMEDIATELY UNDER THE PECTORALS.

I. Genus.—GOBIOIDES.

FISHES of the fourth order, with the two ventral fins united into one, with a fin on the tail, and with only one fin on the back.

The two species of this genus, which I have seen in the Ganges, are pretty common, but are in little request among the natives, especially among the Hindus, most of whom abstain from eating them. They grow to about a foot in length, and have a strong affinity to the *Gobioides anguilliformis* of La Cepede, or the *Gobius anguillaris* of Linnæus. Both, indeed, differ very much from the description of the *Gobius anguillaris* given by Walbaum in his edition of Artedi, (Vol. II. p. 200;) but this excellent ichthyologist confesses, that the fish which he meant was different from that of Linnæus. The differences between our Gangetic fishes and the *Gobioides anguilliformis* of La Cepede (*Hist. des Poissons*, Tome II. p. 577) will be best seen by comparing the descriptions.

Both our Gangetic species have long slender eel-like bodies, stained more or less with red. Their fins are without spots.

Their *heads* are small, blunt, and devoid of scales. Their *mouths* are large, and descend obliquely backward, the under *jaw* being somewhat the longer. Their *eyes* are minute, and on the summit of the head. Their *gill-covers* consist of single plates, are quite entire on the edges, are covered with a thick smooth skin, and move freely. The *membranes* of the gills project under the covers, and leave on each side a moderate *aperture.*

The *lateral line* is in the middle of the side, and is straight. The *vent* is before the middle.

There are six fins. The *dorsal* runs nearly the whole length of the back, and all its rays are nearly equal to each other in length. The *pectoral* fins are very short, and rounded. The *ventral* fin, composed of two united, is concave below, short, and rounded. The *anal* fin extends the whole length of the tail, and its rays are nearly equal in length to each other. The *tail* fin is oval, and ends in a very sharp point.

1st Species.—Gobioides rubicundus. Plate V. Fig. 9.

A *gobioides* with the back, tail, and anal fins united, and with forty-six rays in the first, and forty in the last.

This fish is found in the estuaries of the Ganges. It is of a lurid appearance, and faint *colours;* dirty brown above, and red below.

The *head* is wider than the body, and marked near the *snout* with a kidney-shaped depression. The upper *jaw* in opening is protruded; the under ends in a point turned inwards. In each jaw is a row of long sharp uncovered *teeth,* which project forward. On the inside of these large teeth are many small ones, scattered without order. The *tongue* is smooth, fleshy, blunt, and tied down its whole length. The *palate* is smooth. Each *nostril* consists of one small aperture near the eye. The gill *membrane* contains five rays.

The *throat* projects a little. The *lateral line* is sunk, and sends off veins, as it were, towards each side. There are no *scales.* The *tail* is like a wedge.

The *back* fin contains forty-six rays, each divided into two. The *pectoral* fins are near the middle, and contain many minute rays, which I cannot number. The *ventral* fin contains twelve divided rays. The *anal* fin contains forty rays. The *tail* fin, although united with the dorsal and anal, is readily distinguished from both. The number of its rays I cannot satisfactorily ascertain.

2d Species.—Gobioides ruber.

A *gobioides* with dorsal and anal fins distinct from that of the tail, the former containing fifty-five rays, and the latter fifty-two.

This fish I have found in the estuary below Calcutta. The whole is of a strong but dirty red *colour*.

The *head* is rather narrower than the body. Both *jaws* contain *teeth*, which are concealed by fleshy *lips*; nor, when opened, does the upper protrude. The *nostrils* are not distinguishable. The rays of the gill *membrane* are not to be seen, owing perhaps to its thickness.

The body is covered with small remote *scales*.

Each of the *pectoral* fins has twenty rays. The *ventral* has ten. The rays of the *tail* fin are sixteen.

II. Genus.—Gobius.

Fishes of the fourth order, with the ventral fins united into one, and with two fins on the back.

This genus, of which I have observed twelve species in the Ganges, may be divided into three kinds, differing a good deal in their appearance and manner of life.

1*st*, Such as resemble eels in the form of their bodies.

2*d*, Such as have their pectoral fins inserted into a muscular protuberance, analogous somewhat to the shoulder in quadrupeds.

3d, Such as have short compressed bodies, without the above mentioned protuberance.

Of the FIRST DIVISION I have found three species. They have a very strong affinity to the two fishes last described, and by the natives are considered as belonging to the same genus, which those of Bengal name *Changua;* but the *Changuas* belonging to the genus *Gobius* are not near such ugly looking animals as the *Gobioides*, and are considered as excellent eating by all ranks of the natives. When alarmed, these fishes, like eels, very suddenly disappear in the sand or mud, in which they dig with great facility. I have heard of two other fishes (no doubt species of *gobius)* that have similar manners; but I had no opportunity of examining them, nor of knowing whether they belonged to this division, or to one of the other two. The one is named *Baliya kadi*, and, from its name, should belong to the third division. The other is named *Kadi changua*, and, from its name, should belong to the first. The general characters belonging to this division are as follows:

They are small fishes, from six inches to a foot in *length*, very slender, like eels, and *variegated* with numerous spots or dots.

The *head* is blunt, slopes much from the eyes, and is small. The *mouth* is low in the head, and opens horizontally rather wide. The upper *jaw* is the longest, and does not protrude in opening. Both contain a single row of *teeth* turned forward; and at the end of the under one there is a tubercle containing two teeth larger than the others. The *eyes* are on the summit of the head. The *gill-covers* are smooth, and move freely. The *membrane* is distinctly visible, and its rays may be readily numbered. The *openings* of the gills are of moderate size.

The *lateral line* is in the middle of the side, and straight. The *vent* is before the middle.

There are seven fins. On the *back* are two, of which the hindermost is by much the longer, and its rays are nearly of equal lengths: neither has prickles. The *pectoral* fins are short and round. The *ventral* fin contains twelve rays. The *anal* fin runs almost the whole length of the tail, nearly opposite to the hindermost on the back, and its rays are all nearly of equal length. The *tail* fin is oval, and ends in a sharp point.

These fishes have a strong affinity to the *Gobius lanceolatus* of Bloch, (*Ichth.* Tome II. p. 7, Plate XXXVIII. Fig. 1,) but are abundantly distinct species.

1st Species.—Gobius bato. Plate XXXVII. Fig. 10.

A *gobius* with an eel-like body, with five rays in the first dorsal fin, twenty-one in the second, and twenty-two in the fin behind the vent.

This small fish is found in the estuaries of the Ganges. The body is slippery, above of a pale green *colour*, with numerous black dots, and beneath white, with a silvery gloss on the sides. The whole is somewhat diaphanous, so that the muscles are visible, their divisions forming, as it were, veins which proceed from the lateral lines. The upper fins are dotted. The eyes are golden coloured.

The *mouth* is very large. The upper *lip* is fleshy; the under scarcely observable. The *teeth* are sharp, and many of them have a double point. The *tongue* is blunt, its root is covered with tubercles, and it is of a green colour, with a golden gloss. What I take to be the *nostrils*, are two minute tubercles on the

10

sides of the upper lip. The *eyes* are rather small and globular, but do not project from the head: their pupils are circular. Each of the *gill-covers* consists of two plates: their *membranes* contain three rays.

The *tail* is wedge-shaped. There are no *scales*.

The *dorsal* fins are slightly united. The first is no higher than the second, and is rounded: its rays are undivided, and bent back at their ends; those of the second are split into two, and recurved at the ends, the last being divided to the root. The *pectoral* fins are near the middle, and each contains twenty-one rays. The *ventral* fin forms a kind of oblique hood, is oval, and ends in a sharp point. The rays of the *anal* fin are recurved, and the last is divided to the root. The *tail* fin contains about twenty-two rays, some of which on each side are very short.

2d Species.—Gobius changua. Plate V. Fig. 10.

A *gobius* with an eel-like body, with five rays in the first, and thirty-one in the second back fin, and with twenty-nine in that behind the vent.

This species is found in the estuaries of the Ganges.

The body is slippery, above of an olive *colour*, clouded with black dots, and beneath white. The tail fin is beautifully spotted with black. The eyes are white.

The *head* is devoid of scales. Both *lips* are fleshy: the upper one, on each side, has a sharp projection, perhaps the *nostrils?* The *teeth* are straight, and rather blunt. The *tongue* is blunt and smooth. The *eyes* are small and protuberant, and approach close to each other: their pupils are circular. The *gill-covers* are simple; each of their *membranes* contains four rays.

The *tail* ends in a rounded wedge. The gill-covers and

body are covered with minute remote *scales*, which adhere firmly.

The first *dorsal* fin is rounded, slightly connected with the second, and no higher: its rays are undivided and recurved; those of the second are recurved, and branch at the tops. The *pectoral* fins are situated low, and are of an oval form, each containing eighteen branched rays. The *ventral* fin is of an obliquely hooded form, and oblong shape. The rays of the *anal* fin are branched; the *tail* fin contains eighteen.

3d Species.—Gobius viridis. Plate XXXII. Fig. 12.

A *gobius* with an eel-like body, with five rays in the first, and twenty-six in the second dorsal fin, and with twenty-six in that behind the vent.

This perhaps may be the *Gobius pectenirostris* of La Cepede, (*Hist. des Poissons*, Tome II. p. 542,) but the description which he gives is so short, that I cannot ascertain the point.

This small fish is found in the estuaries of the Ganges.

The *body* is smooth, above of a dark green *colour*, spotted with black, beneath white. The back and tail fins are coloured like the back, and of the former the first, and the latter are edged with black. The eyes are golden coloured.

The *lips* are fleshy: the upper is double, the outer having a sharp projection on each side; while the inner of the upper and the under lips are fringed, as it were, with short tendrils. The *teeth* are remote, blunt, and parallel to each other. The *nostrils* are near the *eyes*, which are quite contiguous to each other, and the animal possesses the power of protruding them very far, and of again drawing them within the level of the head. They are small, and have oval pupils turned towards the side of the head. The *gill-covers* consist of one plate without scales, and each of their *membranes* contains three rays.

The *belly* is rather protuberant, and the *tail* is compressed. There are very minute *scales* on the body and upper part of the head.

The first *dorsal* fin is quite separate from the second, and at least three times higher. It is in form of a wedge, and supported by soft undivided rays, the middle one of which is much the longest, and the two outer ones by much the shortest. They are situated close together, with scarcely any intervening membranes. The rays of the second fin are subdivided at the ends, and the last is divided to the root. The *pectoral* fins are situated low, and are rounded; each contains nineteen rays. The *ventral* fin is plain and oval. The rays of the *anal* fin are divided, the last being split to the root. The *tail* fin is oblique, and contains sixteen rays.

Of the SECOND DIVISION of the genus *gobius*, one species only, the *G. schlosser*, was known to La Cepede; but in the Ganges I have seen five species, all perhaps distinct from that described by the French naturalist. It seems to be of this division of fishes that Pliny writes, (*Hist. Mundi*, Lib. I., Cap. 19,) "Exit in terram—et in Indiæ fluminibus certum genus piscium, ac deinde resilit."

In fact, the fishes of this division are found in the estuaries, and, while the tide is retiring, remain on the mud which the water has just left, and on which they move, with considerable ease, by means of the muscular protuberances in which their pectoral fins are inserted. While on the mud, they have an excellent opportunity of seizing the small aquatic animals that are returning to the river with the last drainings of the tide. At the same time, these fishes, which are quite defenceless against birds, keep always so near the water, that, with one spring, they can throw themselves into it, in which operation, also, they are powerfully assisted by the muscular protuberances above mentioned. The habit which these fishes have of remaining on the wet mud, not only is a means of more readily procuring food, but secures them from numerous fishes of prey, to the attacks of which their defenceless size would expose them; and they are singularly enabled to watch against the attacks of all their enemies, by the structure and position of their eyes, which entirely resemble those of the *Gobius viridis* above described, a fish which has also the habit of passing much of its time in the air, and serves to connect the two first divisions of the genus *gobius*.

Besides the muscular protuberances, and the singular position of the eyes, by which this second division of the genus *gobius* are remarkable, all the species which I have seen agree in the following particulars.

They are small fishes, seldom exceeding three or four inches in length, although occasionally they are found a foot long. They are found in great numbers, but are little sought after by the natives, and never used by the Europeans. Their general *colour* is dull, so as much to resemble that of the mud in which they delight; but their fins, when spread, possess very considerable beauty. As to *form*, the transverse section of their bodies is nearly circular; but, although long in proportion to their thickness, they have nothing of that flexibility which distinguishes eels or serpents. Their tails are compressed, and end in the form of a wedge.

Their *heads* rise up gently from the first back fin to the eyes, and then slope down suddenly to the mouth: they are small in proportion to the animal, and covered with scales. Their *mouths* are large, and situated low in the head, parallel to its lower part. The under *jaw* is the shortest. The *lips* are fleshy. The *teeth* vary in the different species, but all have a single row in each jaw. Each *nostril* has one small aperture close to the eyes. The gill-covers move freely, and expose the membrane to view. The aperture of the gills is of moderate size.

The *vent* is nearly in the middle. There are small *scales* all over the body, and on the greater part of the head.

The *dorsal* fins are quite distinct. The rays of the first are undivided, but not prickly, and are longer than those of the second; all of which are nearly equal in length. In all the species that I have seen, all these rays are separated by distinct wide membranes; in which respect they differ from the *Nettee Kunla Motah* of Russell, (*Indian Fishes*, Vol. 1. No. 54,) a fish nearly allied to them; but, as Dr Russell says, that the rays of the first dorsal fin unite into a setaceous thread, his fish has also a strong affinity to my *Gobius viridis*, although it belongs, no doubt, to this second division of the genus. The *pectoral* fins are situated low, and are rounded. The *ventral* fin always contains twelve rays; but differs in shape in different species. The *anal* fin is placed opposite to the second on the back, and is of a similar form. That of the *tail* is very oblique, and always undivided, ending in a sharp point.

4th Species.—Gobius plinianus. Plate XXXV. Fig. 13.

A *gobius* with the pectoral fins inserted into a muscular protuberance; with five rays longer than the membrane in the first fin of the back; and with twenty-six rays in the second fin of the back, as well as in that behind the vent.

This being the most common species, and all the native names of these fishes being generic, I have named it after Pliny, to commemorate the care which this distinguished Roman bestowed on the investigation of nature.

This fish above is of a brownish *colour*, and beneath whitish, with many irregular and incomplete transverse black belts. All over the head are many little tubercles, which, on the lower parts, are white or bluish. The dorsal fins are beautiful: the first is yellow, with bluish spots, and the second is dusky, with incomplete belts of white and blue, edged with black. The pectoral and tail fins are edged with black.

The *head* is wider than the body. The under *jaw,* which wants a lip, has, in its middle, a protuberance containing two erect *teeth.* On each side of this protuberance are many teeth, parallel and near to each other, and projecting forward. In the upper jaw are about six sharp teeth, remote from each other. The upper *lip* is pendulous, and on each side has a sharp projection. The *tongue* and *palate* are blackish. The gill *membrane* on each side contains three rays.

The *scales* on the head and back are remote from each other, and circular; on the sides and belly they approach near, and are quadrangular; on the tail they are imbricated. There is no *lateral line.*

The first *dorsal* fin is rounded, the two outer rays being the shortest, and the one in the middle the longest. It is more than twice the length of the rays of the second fin; and all exceed the membrane by about a third of their length. The first five rays in the second back fin are undivided, the next seventeen are divided into two, and the last four are more subdivided. The *pectoral* fins are indented on the edge, each containing seventeen bifid rays. The *ventral* fin forms a kind of oblique hood, nearly circular: one ray on each side is simple, the others are compound. The fin of the *tail* has seventeen rays, of which some of the lower are very short.

5th Species.—Gobius septemradiatus.

A *gobius* with the pectoral fins inserted on a muscular protuberance, and with seven short rays in the first dorsal fin, thirteen in the second, and fourteen in that behind the vent.

This comes nearest the *G. schlosseri* of La Cepede, but differs somewhat in the number of rays in the fins; and in other respects his description is so short, that no one can be certain whether or not this and his fish are of the same species.

With respect to *colour*, the black belts, very conspicuous on the fourth species, on this little fish are scarcely observable; and it has very few white or blue spots on its body, nor any tubercles on its scales. The first dorsal fin is black, with a transparent edge; the second is transparent, with two black longitudinal stripes. The pectoral fins are entirely transparent.

The under *jaw*, which is covered by a *lip*, has no protuberance in its centre, and contains a few erect *teeth*, while there are many in the upper jaw. The upper *lip* is double, the outer one having a sharp process on each side, while the inner consists of two lobes. The *tongue* and *palate* are white. Each *membrane* of the gill-covers contains two rays.

There are no *lateral lines*, and all the *scales* are imbricated.

The first *back* fin slopes gradually down from the first ray, which is the longest; but, like the others, scarcely exceeds the membrane in length. The first nine rays of the second fin are undivided, the tenth is bifid, and the three last are branched. In each of the *pectoral* fins are thirteen rays. The *ventral* fin is almost plain. The rays of the *anal* fin are undivided. The *tail* fin has seventeen rays, of which two or three below, and five or six above, are very short.

6th Species.—Gobius novemradiatus. Plate II. Fig. 14.

A *gobius* with the pectoral fins inserted into a muscular

protuberance, with nine rays in the first dorsal fin, fourteen in the second, and fifteen in that behind the vent.

This little fish, like the last, has no tubercles on its scales, and scarcely any traces of the black *coloured* belts by which the *G. plinianus* is distinguished; but it has some white spots on its lower parts. The first back fin is black edged with white, with red membranes between the first three rays, and with some intermixture of white behind. The second fin is white, with two longitudinal black belts. On the centre of each pectoral fin there is a large rusty spot.

The upper *lip* is double, the outer having on each side a sharp projection, and the inner consisting of four lobes. There are many parallel close set *teeth* in the upper jaw, six large teeth in the middle of the under one, and many small ones in its sides. I can observe no rays in the *membranes* of the gill-covers.

The *lateral line* may be traced running straight in the middle of the side.

The first *dorsal* fin slopes gradually down from before, and its first ray is double the length of the membrane, while the others scarcely exceed it. The last ray of the second dorsal fin is divided to the root. Each *pectoral* fin contains thirteen rays. The *ventral* fin is in no way concave below; but consists of two blunt lobes, the innermost ray of each being divided to the root into four. The last ray of the *anal* fin is divided to the root into two. The *tail* fin contains thirteen rays, some of which on each side are short.

7th Species.—Gobius tredecemradiatus.

A *gobius* with the pectoral fins inserted on a muscular protuberance; with thirteen rays in the first, and fourteen in the second back fin; and with twelve in that behind the vent.

This little fish, like the two preceding, wants the tubercles and belts, which distinguish the *G. plinianus.* The back and belly are of a brownish *colour*, the sides blackish, with many small green spots. The back fins are black at the bottom, and blue towards the edges, which are dotted with red. The pectoral fins are olive, with blue membranes. The vent fin is dotted with black.

There is a *tubercle* shaped like a crescent between the eyes and mouth. The *lips* resemble those of the *G. septemradiatus.* There is no tubercle on the middle of the lower *jaw.* The *teeth* in both are strong, erect, and distant from each other. The *tongue* and *palate* are white. Each *membrane* of the gill-covers contains three rays.

All the *scales* are imbricated. The *lateral lines* cannot be traced.

The two *back* fins approach close to each other. The first slopes gradually backward, and its first ray is double the length of the membrane, while the others are no longer. The first ray of the second fin is undivided, the others are bifid, and the two last stand very near each other, and might be reckoned as one. In each *pectoral* fin there are twelve rays. The *ventral* fin has no cavity below, and is very deeply divided into two lobes. The fin behind the *vent* contains twelve simple rays; the *tail* fin has fourteen, besides some very short ones on its under side, which cannot be distinctly numbered.

The THIRD DIVISION of the genus *Gobius* is better known to naturalists than the two former, most of the species that they describe coming under this head. I have found five species in the waters tributary to the Ganges, all of them distinguished by the following circumstances.

They approach in form the *G. niger* of Bloch, (*Ichth.* Tome II. Pl. I. fig. 2, 3, 4, et 5,) the sea gudgeon or pink of the English. Their *bodies* are long in proportion to their thickness, but have nothing of the eel-like flexibility. They are slightly compressed, and their tails are shaped like a wedge. The back slopes gradually and slowly from the first fin to the mouth.

Their *heads* have no scales, and are blunt. They have *teeth* in both jaws. Their *eyes* are placed high up, but not near each other; nor do they project from the head. They are rather small.

The *scales* on the body are rather large in proportion to the size of the fish.

The *back* fins are quite distinct from each other. The *pectoral* fins are inserted directly into the side without the intervention of a muscular protuberance.

8th Species.—Gobius gutum.

A *gobius* with eleven rays in the second back fin, which is nearly as broad as the first; with thirteen rays in each pectoral; and with a diaphanous body clouded with black dots.

This fish I found in the lower parts of the Padda or Padma River, which Major Rennell calls the Great Ganges. It is three or four inches in length, and is neither very common nor much sought after. Like the following, it has a very strong affinity to the *Gobius eleotris*, (La Cepede, *Hist. des Poissons*, Tome II. p. 537.)

The *body* is rough, and of a greenish *colour*, with many black dots, clustered into irregular spots, resembling clouds in form. The fins of the back and tail are spotted.

The *head* is small, and narrower than the body. The *mouth* is situated low, and is large, the under *jaw* being the shortest. The *teeth* are sharp; the *eyes* small, with greenish irides, within which a white circle surrounds the pupil. In the *membrane* of each gill-cover there are three rays.

The *lateral lines* can scarcely be traced. The *scales* adhere firmly, and are rough. The *vent* is before the middle.

The first *back fin* contains six undivided rays, of which the two first are not separated by a membrane. The first ray of the second fin is undivided, and the others branched. The *pectoral* fins are rounded; the *ventral* is circular, and contains ten rays. The fin behind the *vent* has eleven rays, and that of the *tail* has seventeen, and is rounded.

9th Species.—Gobius giuris. Plate XXXIII. Fig. 15.

A *gobius* with ten rays in the second dorsal fin, twenty-two in each pectoral, and nine in the fin behind the vent.

In describing the 8th Species, I have already mentioned the affinity of this to the *G. eleotris*, but that fish has no spots on the body. The *Giuris* has a strong affinity also to the *Koker* of Dr Russell, (Indian Fishes, Vol. I. No. 51,) but differs in having much larger fins, and in the number of rays which they contain.

In all the ponds and fresh water rivers of the Gangetic provinces this is a very common fish, and it is a light well-flavoured food, much in request among the natives. It is from six to twelve inches in length.

The *body* is rough, of a dirty pale-green *colour*, with some gloss of gold on the sides, and rather diaphanous below. On the back and sides there are numerous black dots, forming large irregular spots, shaped somewhat like clouds. The second back fin, and that of the tail, are spotted with black in many rows.

The *head* is rather wider than the body. The *mouth* is very large, and descends backward. The *jaws* are flattened, the under one being the longest, and the upper, in opening the mouth, is protruded. The *teeth* are sharp, distant from each other, recurved, and large, but unequal in size. The *tongue* is smooth, is split in two at the end, and fastened its whole length. The *palate* is smooth. Each *nostril* consists of two apertures placed midway between the eyes and mouth. The *eyes* are far back, and of moderate size. The *gill-covers* move freely, and conceal their *membranes*, each containing five rays, the three middle ones of which are very near each other. The gill *openings* are large.

The *lateral line* runs straight in the middle of the side. The *vent* is near the middle. The *tail* is rounded at the end. The *scales* are imbricated, terminate in an angle behind, and adhere firmly.

The first *back* fin is scarcely higher than the second, and contains six undivided strong rays, a little longer than the membranes, which are interposed between every two: the first is a little shorter than the second, which is the longest; the others gradually diminish in length. The rays of the second fin are nearly of equal lengths, and the fin behind ends in a sharp point. Its first ray is undivided, the others are branched; and the last is divided to the root into two, so that some may choose to say that there are eleven rays. The *pectoral* fins are large and round, and their rays are branching. The *ventral* fin is small, and forms a kind of hood below. It contains twelve branched rays. The fin behind the *vent* resembles the second of the back, and the first and last of its rays are simple, while the others are branched. The *tail fin* is rounded, and contains fourteen branched rays, besides both above and below several short undivided ones, the number of which cannot be accurately ascertained.

10th Species.—Gobius sadanundio.

A *gobius* with eight rays in the hinder fin of the back.

This fish is found in the estuaries near Calcutta, and grows to about the length of a man's finger. The specimens I saw were somewhat putrid.

The *form* tapers to the end of the tail. The *colour* is greenish, spotted with black, and having some white on the belly. The hinder fin of the back, with those behind the vent, and on the tail, are spotted with black.

The *head* is oval, and broader than the body. The *mouth* descends obliquely backward. Each *nostril* has only one aperture.

The edge of the *back* forms an arch. The *lateral lines* are faint. The *vent* is nearly in the middle. The foremost *dorsal* fin slopes backward, and has six undivided rays, of which the third is the longest. The *pectoral* fins are rounded, each containing eighteen rays. The *ventral* fin has nine rays. The *tail* fin is rounded, and contains twenty-three rays, of which three on each side are short, and almost united.

11th Species.—Gobius chuno.

A *gobius* with seven rays in the second fin of the back.

This minute fish, which is from one to one and a half inch in length, is found in the estuary below Calcutta, and is of no value either for quantity or quality.

The *colour* is whitish and diaphanous, with some dots on the head.

The *head* is wide and large. The *mouth* descends backward, the under *jaw* being the longest. The structure of the *teeth* in such a minute animal cannot be readily ascertained, although these organs evidently exist. The *eyes* are far forward.

The *lateral line* is straight. The *scales*, although large, are so thin that they cannot be seen without a magnifying glass. The *vent* is near the middle.

The first fin of the *back* has five rays shorter than those of the second, which slopes down toward the tail. The *pectoral* fins are sharp above, and each contains ten rays. The *ventral* fin contains five, and the fin behind the *vent*, which slopes down towards the tail, has ten rays. The *tail* fin ends in a slight crescent, and contains about eighteen rays.

12th Species.—Gobius nunus.

A *gobius* with five rays on the second fin of the back.

This fish is found along with the former in the river below Calcutta, and is still more minute. It is of a green *colour* on the back, and below shines like silver. It has six irregular black belts; one passing through the eyes, a second on the gill-covers, the third at the pectoral fins, the fourth at the vent fin, the fifth behind the second back fin, and the sixth at the end of the tail.

The *head* is oval and wide. The *mouth* descends backward, the under *jaw* being the longest. The *teeth*, owing to the minuteness of the creature, are scarcely discernible. The *eyes* are far forward.

The *lateral lines* are scarcely distinguishable.

The first *back* fin contains six undivided rays. The *pectoral* fins are sharp-pointed, and each contains ten rays. The *ventral* contains the same number. There are seven in the anal fin; and that of the tail, which is rounded, contains about eighteen.

III. Genus.—Cheilodipterus.

Fishes of the fourth order with the ventral fins distinct, with two fins on the back, and with the gill-covers quite entire and unarmed, leaving a large aperture on each side.

These fishes, perhaps, should have been considered as belonging to the genus *Gobiomorus* of La Cepede, the affinity of the two first species to the last division of the genus *Gobius* being so great, that they are considered by the natives as belonging to the same genus, which they call Bele, or Baliya. The first species has no teeth, and the rays of its first dorsal fin are soft; but the second species, which has teeth and prickles on its back, connects it strongly with the third species, very nearly allied to several species of *Perca.* This, joined to all the species having large openings to the gills, and eyes remote from each other, has induced me to place the whole under the genus Cheilodipterus. Perhaps, however, by following a division of which La Cepede has not availed himself, and by separating the thoracic fishes with prickly fins from those which are unarmed, the first of the three following fishes should be separated from the other two, and should be immediately followed by the genus *Ophiocephalus,* to which it has a considerable affinity.

The circumstances common to these three fishes, besides what is contained in the generic character, are as follows:

Their *length,* in proportion to their thickness, is considerable, and their *bodies* are very slightly compressed. The head, gill-covers, and body, are covered with *scales.* The *eyes* are small and high, but at a distance from each other. There are eight fins, and the two on the back are quite distinct.

1st Species.—CHEILODIPTERUS CULIUS. Plate V. Fig. 16.

A *cheilodipterus* with the tail fin rounded, and with six soft rays in the first fin of the back.

This small fish, which grows to three or four inches in length, is pretty common in the ponds and ditches of Bengal, but is little valued. It is tenacious of life, and is remarkable for the *colours* above being lighter than those on the belly. In dirty stagnant pools, overgrown with weeds, it is, indeed, almost entirely black; but, in cleaner water, the back is ash-coloured. The fins on the back and tail, and behind the vent, are diaphanous, not inelegantly variegated with black spots. There is a very dark black spot on each side, at the insertion of the upper edge of the pectoral fin.

The *tail* ends in form of a wedge.

The *head* is blunt, and forms half an oval, the upper part being flat, and the under convex. It is of a moderate size. The *mouth* descends obliquely backward, and has fleshy *lips*, the inferior being rather the longest. There are scarcely any *teeth*, except a slight roughness in the jaws. Each *nostril* has two apertures, of which the foremost is tubular. The *gill-covers* are dotted, covered with muscles, and move freely. They leave exposed their *membranes*, each of which contains four rays.

The *back* is nearly straight, and the *belly* arched. There are no *lateral lines*. The *vent* is near the middle. The *scales* are imbricated, of a moderate size, adhere firmly, and are dotted, but are quite smooth on the edge.

The rays in the first fin of the *back* are undivided, recurved, and nearly of equal lengths. The second fin is arched, blunt at both ends, and contains ten rays, of which the first is undivided, and the others branched; but the two last approach at the base, and may be considered as one. The *pectoral* fins are near the middle of the body, and rounded: each contains seventeen rays. The *ventral* fins are small and sharp, and each contains six rays. The fin behind the *vent* is opposite to and shaped like the second dorsal, and contains eleven rays, of which the first is very short and undivided, and the others are branched. The *tail* fin contains twenty-seven rays, about six of which on each side are short and undivided.

2d Species.—Cheilodipterus butis.

A *cheilodipterus* with a rounded tail fin ; and with six prickles in the first fin of the back.

This fish is found in the river below Calcutta, but not in great numbers. It does not exceed a span in *length*, and is of a blackish *colour*, with the hinder fins spotted with red, and the ventrals with black. There are two red spots at the roots of the pectoral fins.

The *head*, viewed from above, is blunt, and slopes considerably downwards, so that the *jaws* are flattened, the under one being the longest. In both jaws are small *teeth* almost like bristles. Each *nostril* has only one aperture near the eye. The *eyes* are black. The *gill-covers* end in a point, and in each of their *membranes* there are seven rays.

The *back* is arched, the *belly* is straight. The *lateral lines* are not observable. The *vent* is nearly in the middle. The *tail* is a little compressed, and rounded at the end. The *scales* on the body are large in proportion, and their edges are beset with short, stiff, hair-like incisures.

The *back* fins are near the middle. The first is arched, and its prickles are slender. The second fin is sharp behind, and contains nine rays. There are eighteen in each of the *pectoral* fins, which are short and rounded. The *ventral* fins are rather more forward than the pectorals, are sharp pointed, and each contains five rays. The *anal* fin is on the fore part of the tail, ends sharp behind, and contains nine rays. The tail fin contains eight long rays, and about seven short ones, closely compacted.

3d Species.—Cheilodipterus panijus.

A *cheilodipterus* with the tail fin ending in a straight line ; and with the first fin of the back containing nine prickles, of which the second reaches to the end of the tail.

This fish is pretty common in the estuaries of the Ganges, and affords a rather light and well flavoured diet. It grows to about a foot in *length*, and has a considerable affinity to the *Perca asper* and *Perca zingel*, (Bonnaterre, *Tabl. Encycl.* p. 126, Pl. CXXVI. fig. 206, 207;) but their backs are straight and their bellies protuberant, just the reverse of the case in our Gangetic fish. This is of a greenish *colour*, with a gloss of silver changing to purple.

The *back* slopes from the first prickle towards each end.

The *head* is long and half oval, being flat above, and the *jaws* depressed. The lower part is diaphanous. The upper *jaw*, which is the longest, in opening the mouth, is protruded. The *mouth* is of moderate size and horizontal. The jaws have scarcely any *lips;* but the bones of the upper are pretty large, and at their lower ends terminate as if cut straight across. In both jaws is a single row of small conical *teeth* remote from each other. The *nostrils* are near the eyes, each having two apertures. The *eyes* are yellow, with circular pupils. The *membranes* of the gill-covers are half exposed on the throat, each containing five rays.

The *lateral line* in its fore part is double; but the two branches join towards the end of the tail. The *scales* are of a moderate size, adhere firmly, and are edged with bristly incisures. Near the commencement of the upper branch of the lateral line is a space surrounded by a margin, which is indented like a saw. The *vent* is before the middle.

The first *dorsal* fin slopes backward, and its first prickle is very short, but the second projects far beyond the membrane, and reaches to the end of the tail. The second fin contains twenty-eight rays, of which those in the middle are shorter than those toward each extremity. The *pectoral* fins are short and somewhat sharp, and each contains twenty rays. The *ventral* fins are very short, and each contains six rays, of which the first is a short strong prickle, while the others are branched. The fin behind the *vent* extends along most of the tail, and contains twenty-eight rays. The *tail* fin contains nineteen large rays, besides some short closely compacted ones on each side.

IV. Genus.—OPHIOCEPHALUS.

Fishes of the fourth order with large scales on the head and gill-covers, which are smooth edged and unarmed; with a body very long in proportion to its thickness, and little compressed; with one long unarmed fin on the back; and with two ventral fins.

This genus of Indian fishes, first noticed by Bloch, contains a considerable number of species, all having a very strong affinity to each other, and rather difficult to distinguish by satisfactory characters, owing, in a great measure, to the number of rays in the fins being liable to a considerable variation in different individuals of the same species. The *colours* also vary considerably at different ages, and, I suspect, in different circumstances affecting variously the animal's health, although on this last point I cannot be certain.

All of them, that I know, afford a light and wholesome but rather insipid diet, and are very tenacious of life. In so much, that in China they are often carried alive in pails of water, and slices are cut for sale as wanted, the fish selling dear while it retains life, while what remains after death is considered as of little value.

Besides the characters mentioned as generic, all those fishes, although their *colours* are variegated and sometimes bright, have a lurid appearance. They have the *head* rounded before, and flattened towards the jaws; and it is short in proportion to the body. The *mouth* is large, and descends backward, the under *jaw* being the longest. The *lips* are fleshy, with a small bone belonging to each side of the upper. In both jaws are many small *teeth* crowded together, with some large sharp conical ones intermixed in certain places. The *tongue* is smooth, and not tied down. Each *nostril* has two apertures; one tubular near the jaw, and the other circular near the eye. The *eyes* are of moderate size, and placed at a distance from each other, but rather high. The *gill-covers* move freely, and leave a large *aperture*. Their *membranes* cover the throat, and are there half exposed to view. They are thick; but, although the rays are concealed by the covers, these do not prevent the number of rays which they contain from being ascertained.

The *body* is slippery. When there are *lateral lines* they commence on the shoulders, suddenly bend down behind the pectoral fins, and from thence run along the tail in the middle of its depth. Some of the species, however, have no lateral lines.

There are seven *fins*, of which the *dorsal* extends from the head to near the end of the tail, and that behind the *vent* extends from this organ as far as the dorsal does. Both are very high, and their rays are of nearly equal lengths, but rather longer behind than before. The *pectorals*, which are low down, and the *tail* fin, are rounded. The former are shorter than the head, and the *ventral* fins, which are near each other, are still smaller.

1st Species.—OPHIOCEPHALUS WRAHL? Pl. XXXI. Fig. 17.

An *ophiocephalus* with from forty-three to forty-six rays in the dorsal fin; with many irregular black belts surrounding the body; and with the hinder parts of the back and anal fins spotted with black.

Although La Cepede (*Hist. des Poissons*, Tome III. p. 552) would seem to know this genus merely from the writings of Bloch, yet his description of the colours of the wrahl can by no means be reconciled with that given by Bloch (*Ichth.* Tome X. p. 117, Pl. CCCLIX.) of his *O. striatus*, which La Cepede considers as the same with his *wrahl*. This is a most barbarous name, which, were the two species the same, should, on every account, be rejected for the classical name given by Bloch; but there is reason to believe that the fishes of the two authors are different; for that of Bloch, which I have never seen, has the back of an uniform dusky colour, which, on the sides, is irregularly indented with the white of the belly; while the back and vent fins are obliquely striped, throughout their whole extent, with broad dusky lines. I doubt much of the name *wrahl* being Malay, as La Cepede states, because the missionary John, by whom this name was transmitted to Europe, so far as I know, had no knowledge of the countries where the Malay language prevails; nor has the word any resemblance to the soft sounds of that tongue. After the word wrahl, or what was read as such, John probably put the letters Mal. intended as a contraction for Malabar, the vulgar name by which the Tamul language is called by the Europeans in India. His European correspondents, not aware of this, have probably considered the letters Mal. as a contraction of Malay. I must, however, confess, that, in the dialects of the Tamul, which I heard, as well as in the Bengalese, the fish is called *Sola*, or *Sol;* and *Wrahl* is probably an erroneous reading of what John wrote.

The *wrahl*, or *sol*, is found in the ponds and rivers, both fresh and salt, of every part of India that I have visited, and grows to about two feet in *length*. Its upper parts are of a brownish-green *colour*, variegated with irregular black belts. Below the lateral lines, the sides are variegated with dark and yellow belts, and the belly is white. The hinder extremities of the vent and back fins are yellowish, with many black spots between the rays.

The *head* is broader than the body, with large angular scales, on which are, irregularly scattered, several small circular cavities. On the lower part of the head are scattered many dots, collected into irregular clusters. The upper *jaw* has a notch at the end, and is protruded, when the mouth opens; both have a membranous veil, as it were, on their inside. The *teeth* are imbricated in a double row in each jaw, with many large conical ones intermixed. There are folds on the *palate*. The *gill-covers* end in a point, and each of their *membranes* contains six rays.

The *lateral lines* are plainly distinguishable. The *vent* is rather behind the middle. The *scales* on the body are imbricated, orbicular, large, and smooth on the edge, but rather rough on the surface.

The rays of the *back* fin are all jointed: the foremost are undivided, but nine or ten behind are branched. Each *pectoral* fin contains seventeen dichotomous rays. Each *ventral* fin contains six rays. The *anal* fin contains from twenty-six to twenty-eight rays like those of the back. The *tail* fin contains fourteen long rays, with some short ones on each side.

2d Species.—OPHIOCEPHALUS CHENA.

An *ophiocephalus* with from forty-three to forty-six rays in the back fin; with the back of an uniform colour, and with no spots on the back and anal fins.

This is no doubt the *Muttah* of Dr Russell, (Indian Fishes, Vol. II. No. 162,) which he considers as the same with the *Ophiocephalus striatus* of Bloch; but its colours differ much both from those of that author, and from those of the *sol*, which by La Cepede is considered as the *Ophiocephalus striatus.* I should, however, have taken the *sol* and *chena* for fishes belonging to the same species, had not the fishermen of Goyalpara, where both are very common, considered them as distinct fishes.

Except in *colour* the *chena* differs in nothing essential from the *sol;* and, as it has been described by Dr Russell, I need not enter into any farther detail.

3d Species.—Ophiocephalus lata. Plate XXXIV. Fig. 18.

An *ophiocephalus* with about thirty rays in the back fin, and with many dark belts on the back, descending to a longitudinal dark irregular stripe, which passes through the eye to the tail.

Except in colours this fish differs little from the *O. punctatus* of Bloch, (*Ichth.* Tome X. p. 114, Pl. CCCLVIII.) which name La Cepede (*Hist. des Poissons,* Tome III. p. 554) has most wantonly changed to the barbarous *Karrowey,* that does not differ from *Gorayi,* the name given in Bengal to the young fish, while the adult is called *Lata.* The *colour* of the fish, as described by La Cepede, being of an uniform dirty white, with many small spots, can neither be reconciled to that of the *Lata,* nor to that of the *O. punctatus* of Bloch, which has its back of an uniform black colour; and in the account both of La Cepede and of Bloch the jaws are stated to be of equal lengths, which is not the case in any *ophiocephalus* that I have seen.

The *Lata* of Bengal, in the Tamul language, was called *Mota* by those whom I consulted; but the *Muttah* of the Telingas, evidently the same name, is referred by Dr Russell to the *O. striatus* of Bloch, while the *O. punctatus* of the latter author is referred by Dr Russell to his *sowarah,* (Indian Fishes, Vol.

II. No. 163,) but the tail fin of the *sowarah* ends in a sharp point, which is by no means the case with that of Bloch, nor with any species of *ophiocephalus* that I have seen.

Having premised these difficulties, I now proceed to state, that the *Lata* of Bengal is found in the ponds of all parts of India that I have visited; that, so far as I know, it never exceeds a foot in *length;* and that, for eating, it is considered far inferior to the sol.

The back is of a dirty green *colour*, and the under parts of a pale yellow, the two colours being separated by a broad irregular dark stripe running through the eye to the tail. Across the back, to this stripe, there run many irregular dark belts. On the sides of the adult fish are generally scattered many small black round spots, especially if the water in which the fish has lived is full of dirty weeds. These spots are always wanting in the fish when young. The back and anal fins in the adult are edged with white, and several rows of black spots extend all their length, which is also the case in the young fish, the tail fin also of which is spotted in a similar manner.

The *head* is wider than the body, and is covered with large angular scales, on which are irregularly scattered several small round cavities. In opening the mouth the upper *jaw* is very little protruded, and the bones of the upper lip are minute. In both jaws the fore *teeth* are crowded and small; the hinder ones are in a row, remote from each other, large, and conical. The *tongue* is sharp. The *palate* is smooth behind, but contains many small teeth adjacent to those in the jaw. The *eyes* are far forward. The *gill-covers* end in a point, and each of their *membranes* contains five rays.

The *lateral lines* are with difficulty distinguished from others formed by an elevated line, ending in a raised point, which runs along the centre of most of the scales; but the scales on the true lateral lines end in a notch. The *scales* are large, adhere firmly, are smooth, and have entire edges. The *vent* is before the middle.

The *back* fin is rounded at the posterior end, and its last ray is very deeply divided into two. Each *pectoral* fin contains sixteen rays, of which one on each side is simple, the others are branched. Each *ventral* fin contains six rays, of which the first is undivided, and the others are branched. The *anal* fin is rounded behind, and contains about twenty branched rays: that of the *tail* contains twelve that are long and distinct, besides some short compacted ones at each side.

4th Species.—OPHIOCEPHALUS MARULIUS. Plate XXII. Fig. 19.

An *ophiocephalus* with from fifty-two to fifty-four rays in the back fin, and six in each ventral, and with an eye-like spot on the fin of the tail.

In all parts of India that I have visited, this fish is found in ponds and fresh water rivers; and although often found within the tide-way, so far as I know, it never frequents places where the water is salt. It is found three feet in length; but is not considered so good for eating as the *Sol.* In the lower parts of Bengal, the persons dedicated to religion, from some old prejudice, think it unlucky to say that it is either good or bad.

The adult fish is of a greenish *colour* above, with several dark irregular belts crossing the back, and terminating below the lateral lines in large black irregular spots. The under parts are whitish. There are many white spots scattered over the sides, and on the fins of the back, vent, and tail. At the root of the latter, and near its upper side, is a fine black spot, surrounded, like the pupil of the eye, by a white ring. In the young fish the colours are very different. An orange-coloured stripe runs from the eye to the tail, parallel to the back. The dark belts descend no lower than this, nor are there any white spots on the sides, nor fins. The eye-like spot, however, is always observable on the fin of the tail; but, in the young fish, the black of the centre is surrounded by an orange-coloured ring. The fin of the back, towards its posterior part, has several pale stripes, descending obliquely backward.

The *head* is rather narrow. In opening the upper *jaw* is protruded, and its *lip* on each side is provided with a bone, which is dilated downward. In the whole extent of both jaws there are many small clustered *teeth*, with a single row of larger ones. The *tongue* ends in a sharpish point. The *palate* near the jaws has many small teeth collected in three clusters, the middle one of which, connecting the other two; is in form of a crescent. The *eyes* are far forward, and rather large, and have circular pupils. The *gill-covers* end in a sharp point, and each of their *membranes* contains four rays.

The *lateral lines* are evidently distinguishable. The *vent* is before the middle. The *scales* on the head are very large and angular: on the body they are of moderate size, smooth, and dotted; and, in general, are marked on the centre by a longitudinal elevated line.

The *back* fin is rounded behind, and its rays near the fore part are bifid, while those towards the tail are branched. Each *pectoral* fin contains nineteen rays, of which one on each side is simple, while the intermediate ones are branched. The *ventral* fins are sharp-pointed, and each contains one short undivided ray, and five branched ones. The anal fin is rounded behind, and contains from thirty-one to thirty-five rays. The rays in the fin of the tail are fifteen in number.

5th Species.—OPHIOCEPHALUS BARCA. Plate XXXV. Fig. 20.

An *ophiocephalus* with about fifty-two rays in the back fin, and five in each ventral.

This fish I found in the Brahmaputra river, near Goyalpara, where it inhabits perpendicular banks in holes dug like those of the *Martin, (Hirundo.)* In these it lurks watching for its prey with its head out; and, notwithstanding its strong and variegated colours, is an ugly animal. It is found three feet long, and is considered by the natives as excellent food.

The upper parts are of a dark-green *colour*, the sides yellow, and the under parts white. On the head, back, and sides, are many black irregular spots, with a few red intermixed. Some of those on the head are clustered round a common centre, so as to resemble the rays of a star. The back, tail, and anal fins are yellowish-olive, with numerous black spots, and the two first are edged with red. The pectoral fins are reddish, spotted with black.

The *head* is nearly of the same width with the body. In opening the mouth the upper *jaw* is protruded. The small *teeth* in the upper jaw are not intermixed with larger, but there is a row of the latter in the lower jaw. Immediately behind the jaw there is a line of teeth on the *palate*. The *tongue* is blunt. The *eyes* are far forward, and rather small. The *gill-covers* are blunt, and each of their *membranes* contains four rays.

There is no distinct *lateral line.* The *vent* is before the middle. The *scales* are large and smooth.

The *back* fin terminates behind in a sharp point, and all its rays are divided at the end. Each *pectoral* fin contains sixteen rays, the *anal* has thirty-five, and that of the *tail* nineteen.

6th Species.—Ophiocephalus gachua. Plate XXI. Fig. 21.

An *ophiocephalus* with about thirty-six rays in the fin of the back, and five in each ventral; and with irregular dusky belts across the back.

This fish, which sometimes grows to a foot in length, but seldom exceeds a span, is very common in the ponds and ditches of Bengal, and is one of those which are supposed to fall with the rain from heaven. In fact, with the first heavy showers of the season, it has been often seen leaping and wriggling among the grass; and both by natives and many Europeans is supposed to have fallen with the rain. I have, however, no doubt, that the animal, when thus discovered, has been in search of a more commodious abode. During the dry season it has suffered much from being pent up in half putrid stagnant pools, so that, when the first heavy rains fall, it is eager to enjoy the grateful supply of fresh-water, and wriggles among the moist grass in search of more room, and of the food, which must have been nearly exhausted in the pools that it formerly occupied.

The body above is of a greenish *colour*, and below of a dirty white. Several dark belts run across the back to the lateral lines, descending obliquely forward. The throat and insertions of the pectoral fins are bluish. The fins of the back and tail are greenish, edged first with black, and then with dirty red. The pectorals are pale brick colour, with several rows of blue spots. The anal fin is of a dirty pale bluish-green, edged with black, and then with white.

The *head* is wider than the body. The upper *jaw*, in opening, is little, if at all, protruded. The only long *teeth* are in the fore part of the upper jaw. The *tongue* is rather sharp. The *gums* project behind the teeth, nor has the *palate* any of these organs. The *eyes* are placed forward. The *gill-covers* end in a sharpish point, and each of their *membranes* contains five rays.

The *shoulders* ascend a little from the fin. From the shoulders the *lateral lines* are distinct, and consist on each scale of three small elevated points with a line interposed. The *vent* is rather before the middle. The *scales* are large, imbricated, and striated.

The fin on the *back* is rounded behind, and its rays at the top are divided into two or more branches. Each of the *pectoral* fins contains fifteen divided rays, and each of the *ventrals* five. The *anal* fin contains about twenty-two divided at the tops, and that of the *tail* has about twelve.

7th Species ?—Ophiocephalus aurantiacus. Plate XXIII. Fig. 22.

An *ophiocephalus* with about thirty-four rays in the back fin ; and with the colour of every part of the fish like that of an orange.

This fish I found in a pure mountain stream near Goyalpara, on the N. E. frontier of Bengal. The bottom of the torrent consisted of stones. I have put it down as a distinct species with doubt, because the fishermen gave it the same name with the last described species ; and, on a close examination, I could find no essential differences between the two, except in the colours. These, indeed, are so different, that I can scarcely persuade myself that they originate entirely in a difference of the situation in which the two animals have lived.

The only difference in form that I observe is, that in the *O. aurantiacus* the back does not rise as in the *O. gachua ;* nor is the head so wide in proportion, while the lateral lines cannot be distinctly traced.

The *colours* are, however, most widely different, the whole body and fins of the *O. aurantiacus* being of the colour of orange-peel, with some irregular stains of a redder hue, especially on the sides of the head, and on the pectoral and tail fins.

V. Genus.—Labrus.

Fishes of the fourth order, with the jaws protruded in opening the mouth; with the gill-covers smooth edged and unarmed; with one fin on the back; with a compressed body; and with prickles in some of the fins.

The two following species I refer to the genus *Labrus*, although their almost total want of teeth would, perhaps, require their forming a distinct genus. Notwithstanding the form of their tail fins, they approach nearer the *Labrus malapterus* of Bloch (*Ichth.* Tome IX. p. 26, Pl. CCXCVI. Fig. 2) than to any other fish described by that author. They have so strong an affinity to each other, that, among the fishermen, they are usually called by the same name; but, in the eye of the naturalist, they are abundantly distinct.

1st Species.—Labrus badis. Plate XXV. Fig. 23.

A *labrus* with seventeen recurved simple rays, and eight branched in the back fin; and with three prickles and seven branched rays in that behind the vent.

This small fish, which seldom exceeds three inches in length, is found in the ponds and ditches throughout the Gangetic provinces; but is no where so common, nor is its quality so good as to render it an object of any importance.

The fish is inclined to an oval *form*, but is rather too long to be exactly such, and is much flattened on the sides. The scales at the base shine like silver; in other respects, the colour is subject to much variation. In young fishes, especially those living in clear water, the body is variegated with alternate transverse belts of black and dirty green; but, in old fishes, especially if they have lived among dirty weeds, they are irregularly variegated with black and dirty red, as in the figure. On each shoulder there is always a small round blue spot, surrounded by a black ring. The back fin is blue, variegated with black; and when the fish is very dark, the same variegation extends also to the fins of the tail, and behind the vent, and even to the ventrals.

The *head* is of moderate size and covered with scales, slopes gradually down to the mouth, ends in a blunt form, and is nearly oval. The *mouth* is small, and, when shut, descends almost vertically. The *jaws* contain no *teeth*, unless a little roughness be considered as such. The *tongue* is smooth, and rather sharp. The *palate* is smooth. There are two openings to each *nostril*, the foremost being a short tube. The *eyes*, placed near the middle height of the head, are of moderate size, and convex. The *gill-covers* move freely, end in a point, and are covered with scales. They conceal their *membranes*, each of which contains five rays.

The *back* and *belly* are arched. The *lateral line*, near the end of the back fin, is completely separated into two portions, the foremost being high up and parallel to the back, while the hinder runs straight along the middle of the tail. The *vent* is rather behind the middle. The *scales* are rounded, and finely cut on the edges; in proportion to the fish, they are rather large.

The fin on the *back* reaches from near the head to near the tail, and is rounded behind. The *pectoral* fins are placed low, and are short and rounded, each containing twelve rays. The *ventral* fins are small, and each contains six rays, of which the first is short and undivided. The prickles of the *anal* fin are short, strong, and separated by membranes. The *tail* fin is shaped like a fan, quite entire, and contains sixteen rays, of which one on each side is very short.

2d Species.—LABRUS DARIO.

A *labrus* with fourteen spinous rays, and eight branched ones in the back fin; and with three prickles and seven soft rays in that of the vent.

This fish is found in the ponds and rivers of the northern parts of Bengal and Behar, and is neither larger nor more valuable than the *badis.*

The *colour* in general is silver, with several black transverse belts; but, in dirty water, the black colour extends all over the body. The *form* is rather narrower than in the *badis.*

The *head* is entirely like that of the *badis.* The *mouth* descends obliquely. The under *jaw* is rather the largest: neither contains any observable *teeth.* The *eyes* are high, with the longest axis of the pupils horizontal. The *gill-covers* are rounded and scaly, and move freely, covering entirely their *membranes,* the rays of which I could not exactly number.

Both *back* and *belly* are arched. The *vent* is behind the middle. There are no *lateral lines.* The *scales* are small.

The *back* fin ends behind in a sharp point, and extends almost the whole length of the back. Its prickles are slender. The *pectoral* fins are low, and rounded: each contains nine rays, and each *ventral* six. The *anal* fin is sharp behind. The *tail* fin is circular; the number of its rays I could not ascertain.

I am now to treat of fishes, in arranging which all naturalists have felt the utmost difficulty; nor, so far as I can judge, does

La Cepede's arrangement appear either natural or easy; that is to say, the species included under each genus have no strong general resemblance to distinguish them from those included in other genera; and there occurs the most puzzling difficulty in determining what fishes have only one fin on the back from those having two, and those which have the bones of the head armed with spines from those which have them merely indented on the edges,—the toothlets on the edges, in many cases, being in fact strong spines, while all intermediate degrees of strength may be observed, until, in many species, the toothlets are very short and blunt. I shall here, therefore, leave La Cepede as a guide, and lest a misapplication of the names given by him or Bloch might mislead, I prefer a division of these fishes into three genera, adopted pretty generally by the Bengalese fishermen, adding to each genus a few species, which these men consider as distinct, although they have a strong resemblance to the fishes with which I have classed them.

All the fishes to which I here allude, in both jaws, have either strong conical distinct sharp teeth, or many small sharp teeth crowded together; some of the bones of their heads are distinctly indented on the edges; some of their fins are armed with prickles; and each of their ventral fins contains six rays.

VI. Genus.—Bola.

Fishes of the fourth order with strong distinct sharp teeth in each jaw; with some of the bones in the head indented on the edge; with prickles in some of the fins, and six rays in each ventral; and with opaque bodies compressed to a ridge on the back, but widening gradually towards the belly, so that a transverse section at the shoulders is somewhat triangular.

The fishes of this kind afford a very light, easy digested aliment, but rather insipid; and, from these qualities, especially when of a similar size, they are usually called whitings by the English of Bengal; although they have no other affinity to that fish, the *Gadus merlangus* of naturalists.

Besides the circumstances mentioned in the generic character, the fishes of this genus, which I have observed in the Ganges, agree in the following particulars.

In proportion to either width or thickness, their form is long, and their tail tapers much, gradually above, but rather suddenly below, and is of considerable length. Their backs are arched, and their bellies nearly straight. The body, head, and gill-covers, have large *scales*, which, on the body at least, are finely indented on the edge, and communicate a rough feel. The *vent* is behind the middle.

The *head* is oval and blunt. The *jaws*, in opening, are protruded, and are covered with fleshy *lips*, the upper of which is provided on each side with a moderate-sized bone, which widens considerably downward. The *tongue* is not tied down, and is blunt and smooth. On the *palate* are some naked bones, and some of them contain *teeth*. Each of the *nostrils* has two apertures without tubes. The *eyes* are high, but distant from each other, and are of a good size. The *gill-covers* move freely, and leave large *apertures;* and a great part, if not the whole, of their *membranes* is visible, each containing seven rays, that are readily distinguishable.

The *lateral lines* run parallel to the back, and in their fore parts are high, but behind they pass along the middle of the tail.

The two fins of the *back* are united, but the division is very distinctly marked. The foremost consists of slender prickles, rather longer than the membranes by which they are separated, and growing gradually shorter from either, the first, second, or third, one of which is the longest. The second fin is longer than the first, although not so high in general as the longest of its rays. It commences with one prickle, longer than the last of the first fin, but shorter than the following rays, which are branched, and grow a little longer toward the tail, where the fin is rounded. The *pectoral* fins are low and shorter than the head. The first ray of each *ventral* is either simple and rather hard, or is a slender prickle; and the others are branched. The fin behind the *vent* is much shorter than the second of the back, and is situated forward on the part where the tail suddenly contracts. It contains two or three prickles closely united without any intervening membrane, and several branched rays forming an arch. The *tail* fin is entire, and approaches more or less to the form of a lozenge.

1st Species.—BOLA COITOR. Plate XXVII. Fig. 24.

A *bola* with ten prickles in the first back fin, and one with twenty-six or twenty-seven branched rays in the second; with two prickles and six or seven branched rays in the fin behind the vent; with the first branch of the second ray in each ventral fin very long, and in form of a bristle; and with the snout projecting beyond the mouth.

This fish is found in the Ganges, from the sea up as far at least as Kanpur, and in the Jumna (Yamuna) as far as Agra. It is, however, much more common where the tide reaches; although its quality improves in the upper parts of the river, especially where the shores are rocky. It seldom exceeds a foot in length, and a span is its more usual size. The fish, which Dr Russell has described in his Indian Fishes, (Vol. II. No. 110,) but to which he has given no name, has a considerable affinity to the *Coitor;* but differs somewhat in the number of rays in the fins, and in the form of that of the tail.

The general *colour* is silver, with a greenish brown, changing to purple and gold on the back. The fins of the back and vent are dotted and edged with black. On each side behind the pectoral fin there is an irregular brown spot. The sides are indistinctly marked with narrow dotted lines, ascending upward behind.

The *head* is of a moderate size. The *mouth*, which is large, descends in a curve. The under *jaw* is the shorter. The upper *lip* is double with the outer divided into five. There are many small *teeth* crowded among the row of large ones in each jaw. The *palate* is smooth, and near the throat has two bones covered with teeth. The hinder openings of the *nostrils* are oblong, and near the eyes; the fore ones are at some distance, and are round. The *eyes* are rather large and plain, with their longest axis horizontal, while that of the pupil is vertical. In the *gill-covers* on each side are four bones; the foremost of which is indented with short blunt teeth, the intermediate ones are very small, and the hinder one ends in a sinus bounded by two sharp points. The bones of the *gills* have on the inside a row of tubercles covered with little teeth.

The *lateral lines* are smooth, and may be traced to the end of the tail fin. The *scales* of the head are entire on the edge.

The first prickle of the foremost *back fin* is very short, and joined to the second, which is the longest, without any intermediate membrane. The fourteen or fifteen rays next the prickle, in the hinder back fin, are bifid at the top, and those remaining are branched, the last being divided to the root into two. The *pectoral* fins, each having seventeen rays, are covered at the root with scales, and are rather blunt, but the upper rays are longer than the lower. The *ventral* fins are small, and near each other. Their prickle is short and slender, and the bristle-like division of their second ray is recurved

1

at the top. The fin behind the *vent* is scaly, and contains two strong prickles, the first of which is very short, while the second is shorter than the branched rays, which are six or seven in number, the last being divided to the root into two. The fin of the *tail* ends in a sharp point, is covered with scales, and contains seventeen distinct rays, with some short ones compacted on each side.

2d Species.—Bola chaptis. Plate X. Fig. 25.

A *bola* with ten prickles on the first fin of the back, and one prickle with twenty-five soft rays in the second; with two prickles and seven soft rays in the fin behind the vent; with the first branch of the second ray in each ventral fin lengthened in form of a bristle; and with the lower jaw extending to the extremity of the head.

This fish has some affinity to the *Bodianus stellifer* of Bloch, (*Ichth.* Tome VII. p. 41, Pl. CCXXXI. fig. 1;) but in that species the two back fins are not near so well separated as in ours, while it wants the indented edges of the gill-covers, and its head is armed with several spines. Our fish should rather be a *Centropomus*, according to La Cepede, than a *Bodianus*. It is found in the estuaries of the Ganges, seldom exceeds a foot in length, and is reckoned but poor eating.

The *Chaptis* is silver *coloured* below, while its back is green, with a purple gloss, but without any intermixture of gold. The back and tail fins are dotted and edged with black. On the sides are many lines ascending behind, and consisting of a small elevation on the centre of each scale.

The *head* is rather small and narrow. The two *jaws* are nearly equal in length, and reach to the extremity of the head, extending far backward in a straight horizontal line. In both jaws are some sharp long *teeth* intermixed with some small ones. In the fore part of the *palate* are four naked bones. The *nostrils* are near the *eyes*, which stand far forward, are large, and have their longest axis horizontal, while the pupil is circular. The *gill-covers* contain each four plates, of which the foremost is indented round the edges; the two intermediate ones are very small; and the hinder one ends in a sinus, with a blunt point on each side.

The *lateral lines* may be traced to the end of the tail fin, and are smooth. The *scales* are rather small.

The first prickle of the first *back fin* is very short, as is also the prickle of the second fin. The *pectoral* fins are blunt, although the upper rays are the longest. Each contains nineteen rays, of which the two first and the last are simple, and the intermediate ones are branched. The *ventral* fins entirely resemble those of the *Coitor*. The prickles of the fin behind the vent are strong. The *tail* fin is blunt, and contains seventeen distinct rays, besides some short compacted ones at each side.

What I consider as a mere variety of the above mentioned fish is called *Nuria* by the fishermen of the Yasor (Jessore, Rennell) district. It grows to three feet in length; but on carefully comparing it with the above description, I see no other difference, except that it has one ray less in the hinder fin of the back, (having one prickle and twenty-four soft rays,) which, in this genus, is of very little consequence, the number being liable to more considerable differences in individuals, allowed by all to belong to the same species.

3d Species.—Bola coibor.

A *bola* with nine prickles in the first dorsal fin, and one prickle, with twenty-four soft rays in the second; with three prickles and seven soft rays in that behind the vent; with no bristle on the ventral fins; and with all the lower parts yellow.

This is a very beautiful fish, found in the larger estuaries of the Ganges. I saw only one specimen, which was four feet in length; but it is said to grow considerably larger. At the time, I omitted to take a full description, and only noted the differences between the *coibor* and the description of the *pama*, which follows; and I had no opportunity afterwards of supplying the omission. It may, perhaps, be the *Lutianus chinensis* of La Cepede, (*Hist. des Poissons*, Tome IV. p. 255,) although it ought more properly to belong to his genus *Centropomus*.

The *coibor*, in proportion to its length, is of a thicker *form*, and it tapers less at the tail than the *pama*. The *colour* on the back is similar, but the sides shine like bright gold, and the belly, and lower and tail fins, are of a deep yellow.

The fore aperture of each *nostril* is very small; and behind the row of large tusks in the upper jaw there is a belt of small compacted *teeth*. There are several strong teeth on the inside of the bones, which support the *gills*.

The first prickle of the fin behind the *vent* is minute.

4th Species.—Bola pama. Plate XXXII. Fig. 26.

A *bola* with nine prickles in the first fin of the back, and one prickle, and from forty-one to forty-five soft rays in the second; with two prickles and seven soft rays in the fin behind the vent; with no bristle on the ventral fins; with both jaws extending to very near the end of the snout; and with the sides silvery.

At Calcutta, this fish, when from twelve to fifteen inches in length, is that most commonly called *Whiting*, and is a very light wholesome food, if eaten soon after it is caught; but it soon becomes soft. It sometimes grows to a large size, and is found between four and five feet long. It is caught in great numbers in all the estuaries of the Ganges, but does not ascend higher than the tide.

The *colour* of the body is entirely similar to that of the *Coïtor*. The back and tail fins are dotted, and the first of the former is edged with black. When the fish dies all the extremities soon becom eyellow.

The *head* is rather deep, but not wide. The *mouth* extends to very near the end of the snout, but still is situated low, and is arched backward almost in a horizontal direction. The *jaws* are nearly equal in length. On the inside of the under one is a ridge. In each jaw is a single row of large, crooked, conical, sharp *teeth*, of unequal sizes. In the fore part of the *palate* are four naked bones, and at the throat it is beset with teeth. The *eyes* are lengthened out in a horizontal direction, with the greatest length of the pupils vertical; and are rather small. Both apertures of each *nostril* are circular, and are near the eye. The *gill-covers* consist of three plates, and end behind in one point. The foremost plate, at its angle, is indented with remote incisures. The inferior one is very small. The bones supporting the *gills* along their inner sides have a double row of denticulated projections.

The *lateral lines* are covered with many small scales, and may be traced to the end of the tail fin. The scales on the head are long, and of various sizes.

The first prickle in the foremost *back fin* is rather longer than the second, and is separated from it by a membrane. The prickle of the second fin is short: the first soft ray is undivided, and the others are branched. The *pectoral* fins are sharp above, and each contains seventeen rays. The *ventral* fins are small, and near each other. They want the bristle-like division of the second ray by which the *Coïtor* is distinguished. The fin behind the *vent* is slightly arched at the end, and its prickles are short, the first soft ray being undivided, while the others are branched. The *tail fin* is sharp-pointed, and contains seventeen distinct rays, besides some on each side closely compacted together.

5th Species.—Bola cuja. Plate XII. Fig. 27.

A *bola* with ten prickles in the first back fin, and one prickle with twenty-eight soft rays in the second; with two prickles and seven soft rays in the fin behind the vent; with the first branch of the second ray in the ventral fins lengthened like a bristle; with the under jaw rather the longer; and with the sides variegated longitudinally with black interrupted stripes.

This is a large fish, growing to between four and five feet in length, and is found in the estuaries of the Ganges. By the fishermen it is not classed with the *Bolas*, but with the *Vakti*, a fish which will be described among those belonging to the next genus. It, however, belongs to the genus *Bola*, as I have defined it, while the other *Vakti* belongs rather to the genus *Coius.*

The *Cuja* is in *colour* a very beautiful fish, shining like silver, with a slight tinge of green on the back, while its sides are finely variegated with black spots, disposed in longitudinal stripes. The back fins are finely spotted with black.

The *head* is rather narrow, and slopes suddenly at the eyes. The *mouth* extends to the very extremity of the head, the under *jaw* being rather the longest, and descends a little backward. The upper *lip* is double. In each jaw is a single row of sharp conical *teeth*, nearly straight, and equal in size. In the fore part of the palate are four oblong naked bones. In the throat are two circular ones covered with teeth. The *nostrils* are near the eyes, the hinder openings being oval, and the fore ones circular. The *eyes* are round, and of moderate size, and have circular pupils. The *cheek-bones* and spaces on the nape are indented round the edges as if with short bristles. The fore *gill-covers* are indented on the edges with wider incisures, while the hinder plate terminates in a sinus, bounded by two points. The bones supporting the *gills* are lined with a single row of tubercles, of which some are cylindrical and smooth, and others conical and indented.

Although the ridge of the *back* is sharp, and the edge of the *belly* is rounded, the *sides* do not slope outward so much as in the fishes of this genus before described; still the transverse section is triangular. The *lateral lines* may be traced to the end of the tail fin. The *scales* fall easily off. Above each ventral fin is a long scale-like *appendant.*

The prickles of the *back fin* are strongish; the first is very short, and the third is the longest. The prickle of the second fin is short; the first soft ray is undivided, and the others are branched. The *pectoral* fins are sharp above, and each contains seventeen rays, of which the first and two last are undivided, and the intermediate ones are branched. The *ventral* fins entirely resemble those of the *coitor*, but have stronger prickles. The *anal* fin slopes gradually from before backward; its prickles are very strong, and its soft rays branched. The *tail* fin is blunt, and contains seventeen distinct rays, besides some short ones, closely compacted at each side.

6th Species.—Bola ? coioides.

A *bola?* with eleven prickles and fifteen soft rays in the back fins, which are united; with three prickles and eight soft rays in the fin behind the vent; and with the body spotted with round red and black spots.

This fish agrees so well with the description of the *Medinawa bontoo* of Dr Russell, (Indian Fishes, Vol. II. No. 128,) that I do not think them different species, although Dr Russell says, that the *Madinawa bontoo* never reaches a great size. This I attribute to the large Gangetic estuaries being the proper abode of the fish, and to their affording it a full nourishment, while its growth may be stinted in the ocean, where it was observed by Dr Russell.

The different fishes called *Jacob Evertsen* by the Dutch seamen, and *Jew-fish* by the English of the West Indies, have a great affinity to that of which I am now treating; but all, I believe, belong either to the genus called *Sparus*, or to that called *Bodianus;* while this should be a *Holocentrus*, as that genus is defined by Bloch. I am inclined, however, to believe, that many of the circumstances related by this author (*Ichth.* Tome VII. p. 26, Pl. CCXXIV.) concerning his *Bodianus guttatus*, should be actually referred to the fish I am now describing; although the figure which he gives probably represents another fish.

I have referred this fish to the genus *Bola* with a mark of doubt, because, in the description, which I took long before I had the present arrangement in view, I have omitted to notice whether it possessed the characters by which the *Bola* is distinguished from the following genus called *Coïus*. Its back fins no doubt resemble those of the latter genus; but the essential character has not been placed in these; while the native fishermen place it among the *Bolas;* and although, as will appear from the following description, it has very numerous characters in common with this genus, yet its affinity with the genus *Coïus* is so strong, that this term is applied by these people as its specific appellation, making the whole name, as pronounced by them, *Coïbola* or *Coïbor*.

The *Coïbola* is found in the larger salt-water estuaries of the Ganges, is usually between four and five feet in *length*, is an excellent fish for the table, and possessed of great beauty. The *colour* on the back and sides approaches to olive, and on the belly is white. It has large dusky irregular marks like clouds; and besides many round well defined spots, partly black, partly rust colour, are scattered on the surface.

The *head* is large and wider than the body, and ends in depressed *jaws*, of which the under one is the longer. The *mouth* is very large, and descends backward in a straight line. The bones of the upper *lip* are very blunt, as if cut transversely across. In both jaws, and in the adjacent parts of the palate, are many crowded sharp *teeth*, straight, but of unequal lengths. Some of these, toward the fore part of the jaws, are considerably longer than the others, and are bent backward. On the *palate*, at the throat, there are two round bones. The openings to each *nostril* are both near the eyes; the foremost being small. The *eyes* are far forward, and are round with oblong pupils. The *cheek-bones* end behind in a straight line, and round their edges are indented with toothlets, some of which at the angle are long. Each *gill-cover* consists of three plates, of which the hindmost terminates in three broad and very strong spines. The bones supporting the outer *gills* are covered on the outside with teeth like a comb; those farther in are covered with a row of tubercles.

The fish in *form* is very thick near the shoulders, but slopes toward both extremities; and the *tail* is very slender; but I have omitted to state the form of its *sides*.

There is properly only one *back fin*, the two of which it consists being fully united, although they may be distinguished, as the first is nearly of the same height throughout, while the second rises higher in form of an arch: the first of its rays is undivided, and the others branched, the last being divided to the root. The *pectoral* fins are low, short, and blunt, each containing eighteen rays, of which the first and last are undivided. The *ventral* fins are near each other, and the prickle in each is strong. The *fin* behind the vent is rounded, and its prickles are strong, while the first of its soft rays alone is undivided, and the last is divided to the root into two branches. The *tail* fin is rounded, and contains sixteen distinct rays, besides some short compacted ones at each side.

VII. Genus.—Coïus.

Fishes of the fourth order with strong distinct sharp teeth in both jaws, or with teeth disposed in many rows, with some of the bones of the head distinctly indented on the edges, with prickles in some of the fins, and with opaque bodies, the sides of which are nearly vertical, so that a transverse section is either elliptical, or in form of a lens equally convex on both sides.

The fishes of this kind also afford food of easy digestion, and in general more sapid than that of the preceding genus.

Besides the circumstances mentioned in the generic character, most of the species which I have observed in the Ganges agree in the following particulars; and every species below described, where it is not otherwise specified, possesses these characters.

These fishes are not so *long* in proportion as the Bolas, and their tails taper as suddenly above as below. Their *heads*, at least in part, and their gill-covers and bodies, are covered with *scales*. The *vent* is behind the middle.

The *head* slopes gradually from the commencement of the back fin to the jaws, which are protruded in opening the *mouth*. This is rather large, and has fleshy *lips*, the upper of which has a bone on each side. Each *nostril* has two apertures placed far back. The *eyes* are high, but at a distance from each other, and are of a good size. The *gill-covers* move freely, and leave large *openings*.

The *lateral lines* are high, and arched in a line parallel with the back.

In the species that are well defined, although the two *fins* of the *back*, strictly speaking, are united into one, they may be readily distinguished, by the second being higher than the first, and it is shorter, occupying only the sudden slope of the tail, while it consists of soft rays, and those of the first fin are prickles. In some species, however, the back fins resemble those of the *Pamas*, being more distinct; but the second, in proportion to the first, is always shorter than in that genus. The *pectoral* fins are low. Each *ventral* fin contains six rays, of which the first is a prickle, and the others are branched. The fin behind the *vent* contains some prickles, separated by membranes, and generally more numerous soft branched rays. In several species, both back and anal fins are capable of being received in a *furrow*, lined on each side with scales; and the edges of the hinder fin of the back, and of that behind the vent, owing to their being placed on the slopes of the tail, are nearly vertical.

As I ended the last genus with some ill-defined species, approaching somewhat to this, so I shall commence the account of this genus with some species that resemble the *Bolas*.

1st Species.—Coius vacti. Plate XVI. Fig. 28.

A *Coius* with the tail fin shaped like a fan, with two back fins nearly distinct, the first containing seven prickles, and the second one prickle and eleven soft rays; and with three prickles and eight soft rays in the fin behind the vent.

This fish, by the vulgar English of Calcutta, is called *Cockup*, and is one of the lightest and most esteemed foods brought to table in that city. It has no doubt the utmost affinity to the *Pandoo menoo* of Russell, (Indian Fishes, Vol. II. No. 131,) which he supposes to be the *Cockup* of Calcutta; but, if his delineation of the *Pandoo menoo* be correct, the two fishes are abundantly distinct, as it belongs to La Cepede's genus Perca, while the *Vacti* of Calcutta is a Centropomus. Besides, the *Pandoo menoo* would appear to be a fish very inferior in quality to the *Vacti.*

The *Vacti* abounds in all the mouths of the Ganges, which it ascends as far as the tide, and follows this into marshes. ditches, and ponds; but those found in salt-water are of by far the best quality, as are those about two feet in length. It is often caught five feet long; but, when it approaches this size, the taste becomes strong, and, when small, it is rather insipid.

The *form* is rather thick from the back downward; but it is much compressed at the sides, which are nearly vertical. The upper parts are of a green *colour*, with a gloss of gold and purple, and have some ill defined longitudinal dotted lines, not, however, to be distinguished in all lights. The lower parts are silvery. The fins are pale dusky. The eyes are white, with amethyst-coloured pupils.

The *head* is rather sharp, and its slope forms a curve at the eyes. The space before the eyes is free of scales. The *cheek-bones* are indented on the lower edge. The *mouth* descends obliquely backwards. The *jaws* are within lined by a membranous veil: the under one is the longest, the upper one terminates in a notch. The bones of the upper *lip* widen much below. There are numerous small sharp *teeth* crowded on both jaws, and also on the tongue and palate. The *tongue* is sharp and free. The *palate* has two bones at the throat, and three behind the jaw, all covered with teeth. Of the latter bones the central is in form of a crescent, and the lateral ones are linear. The foremost opening of each *nostril* is circular, the hinder one is vertically oblong. Each *gill-cover* consists of three plates, of which the foremost is indented on the hinder edge, and the hindmost ends in a sharp point. Each of their *membranes* contains seven rays. The bones supporting the *gills* are covered with prickles.

The *back* rises high, and the *belly* is nearly straight. On each side of the nape is an oblong space, surrounded by a broad smooth margin. The scales on the *lateral lines* end in a notch. The *scales* are quite entire on the edges, in which this fish differs from all those nearly allied to it that I have described. On each side above the ventral fin is a long scale-like *appendant*.

The *back* fins are slightly united. The prickles of the first are strong, all separated by membranes, and a little longer than these, the edge of each membrane being concave. The two first prickles are very short, the third is the longest, and the others diminish gradually in length. The second back fin is scarcely so high as the first, extends along the back some way before it suddenly slopes, and forms an arch. The *pectoral* fins are short and rounded, and each contains seventeen rays. The *ventral* fins are of moderate size, and the prickle in each is very strong. The fin behind the *vent* forms an arch; that of the *tail* is scaly, and contains about eighteen rays.

2d Species.—Coius datnia. Plate IX. Fig. 29.

A *coius* with the tail fin bifid; with twelve prickles and ten soft rays in the united fins of the back, and three prickles and eight soft rays in that behind the vent; and with the first branch of the second ray in each ventral fin produced into a bristle.

Some fishermen reckon this a species of *Coius*, others call it a *Vacti*, and it has no doubt a considerable affinity to the last described species. Others, however, consider it as forming a distinct genus, and I have adopted the name which these give it. The *Calamara* of Russell (Indian Fishes, Vol. I. No. 92) has a very great affinity to our *Datnia;* but has a thicker body, and wants the bristle-like elongations in the ventral fins, and the spot on the fin behind the vent.

This is a very beautiful fish, from six to ten inches in length. It is found in all the mouths of the Ganges, and is common in the Calcutta market; but is not near so good for eating as the *vacti*.

The *datnia*, in proportion to its *length*, is a thick fish, and its general *colour* is silver, with faint interrupted longitudinal stripes, which above are green, with a gloss of gold, and below pearly. The first back fin is edged, and the second and tail fin are stained with black. On the fin behind the vent there is a large irregular black spot.

The *head* and back descend in the segment of a circle. The former is oval, somewhat sharp, of a moderate size, and covered with scales to the eyes, before which it is naked. The *mouth* is of a moderate size and terminal, and passes almost directly backward. The under *jaw* is rather the shorter. The bones of the upper *lip* are minute. The anterior *teeth* in both jaws are sharp, conical, of unequal size, and placed in a single row: the posterior ones are blunt and crowded together. The *tongue* is sharp, and slightly tied down. The *palate* is smooth, and, behind the teeth, has a short membranous veil. The posterior aperture of each *nostril* is oblong; the anterior is provided with a short tube. The *eyes* are rather large. Each *gill-cover* contains three plates, of which the foremost, together with the *cheek-bone*, is indented on the edge; and the last terminates in a very strong sharp point. Each of their *membranes* contains six rays: they are concealed by their covers, and are of a thick substance.

The *back* is much arched, and has a furrow for receiving the fin; and the *belly* is rather straight. The *sides* are much compressed. The *scales* are large, indented on the edge, and adhere firmly. Above each ventral fin is a long scale-like *appendant.*

All the prickles of the *back fin* are strong and separated by membranes, the fore part of each of which is attached to the point of the prickle, while the hinder part is attached considerably below the point, rendering the edge of the fin like that of a saw. The first soft ray of the back fin is undivided, and the last is divided to the root into two. The *pectoral* fins are longer than the head, and sharp above: each contains fifteen rays, of which the first and second are short and undivided. The *ventral* fins are rather short, and their prickle is strong: the bristles of their second rays are recurved. The *vent* fin is composed of two fully united; but the first is supported by strong prickles, of which the middle one is the longest; the second contains soft branched rays, the first of which is longer than the third prickle, while the others gradually diminish in length. The *tail* fin consists of two short blunt lobes, of which the under one is the shortest. It contains seventeen distinct rays, besides some short compacted ones at each side.

3d Species.—Coius catus. Plate XXXVIII. Fig. 30.

A *coius* with the tail fin ending in a slight concavity, with the back fins united, and containing ten prickles and fourteen soft rays; with three prickles and eight soft rays in the fin behind the vent; with the first branch of the second ray in each ventral fin lengthened into a bristle; and with a black diffuse spot on each side.

This fish is very nearly allied to the *Doondiawah* of Dr Russell, (*Indian Fishes,* Vol. I. No. 97,) which has twenty-five rays in the dorsal fins. It is also nearly allied to the *Mungi mupudee* of Dr Russell, (*Indian Fishes,* Vol. II. No. 110,) which has five bluish longitudinal stripes, and several rows of crowded teeth in each jaw.

The *Anthias Johnii* of Bloch, (*Ichth.* Tome IX. p. 97, Planche CCCXVIII.) or *Lutianus Johnii* of La Cepede, (*Hist. des Poissons,* Tome IV. p. 235,) is also very nearly allied to our fish, and differs chiefly in all its head being covered with scales, and in the form of the back and ventral fins.

The *catus* is as large as the vacti, but is inferior in quality for eating. It is found in the larger estuaries of the Ganges, and is a very beautiful fish in *colour,* shining with a silvery gloss, and being striped longitudinally on the sides with many rows of black spots, each occupying the middle of a scale. There are, besides, on the back three irregular dark marks, of which the two foremost are indistinct; but the last, placed below the hinder prickles of the back fin, is pretty well defined. The hinder fins are edged, or tinged, with pale red.

The *head* is sharpish, and narrower than the body, but rather large, and of half an oval form. Before the eyes there are no scales. The *mouth* passes straight back parallel to the under part of the head. The under *jaw* is rather the shorter. The bones of the upper *lip* are rather large, and end as if cut transversely. In both jaws is a single row of sharp conical unequal *teeth.* The *tongue* is blunt and free. The *palate,* on its fore part, has three bones, of which the two foremost are smooth. and the other covered with very minute teeth. The

The bones at the entrance of the throat are also covered with teeth. Both apertures of each *nostril* are oblong. The *eyes* are large. The bones of the *cheek* are indented round the edge, with some of the toothlets at the angle rather long. In each *gill-cover* are three plates quite entire on the edge; and the posterior one ends in a point. The *membrane* of each gill-cover is not concealed, and contains seven rays.

The *back* is much arched, and has a furrow for receiving the foremost fin. The *belly* is nearly straight. The edge of the *scales* is minutely indented. The scales are large, and separate easily. Above the commencement of each lateral line is a scaly space surrounded by an indented margin. Above each ventral fin is a short scale-like *appendant*.

The two dorsal *fins*, although fully united, are easily distinguishable, each forming an arch. The edge of the first, as in the preceding species, resembles that of a saw. It has strong prickles, all separated by membranes, and the fifth is the longest. The second fin is scaly at the base, and the first of its rays, all of which are soft, is undivided, and the last is divided to the root into two. The *pectoral* fins are not so long as the head, and are very sharp above. Each contains *seventeen* rays, of which the first, second, and seventeenth, are undivided, and the first and last are very short. The *ventral* fins are near each other, and are rather short. In form they entirely resemble those of the *Datnia*. The fin behind the *vent* is arched, its prickles are strong, and the first of its soft rays is undivided. The *tail* fin ends in a small concavity, and is blunt at both extremities. It has seventeen distinct rays, besides some short compacted ones at each side.

4th Species.—Coius trivittatus.

A *coius* with the tail fin bifid, with eleven prickles and ten soft rays in the dorsal fins; with three prickles and eight soft rays in the fin behind the vent; and with three longitudinal black stripes on each side.

The affinity of this fish with the last is so strong, that, by the natives, both are called by names radically the same, although, as species, they are very distinct. This kind also is found in the mouths of the Ganges, but seldom exceeds six inches in length.

The fish is of a long *form*, flattened at the sides, scaly, and rough to the feel. Except the stripes, it is of a silvery *colour*. The dorsal fins are marked with several black stains, and that of the tail with some longitudinal stripes. The eyes are yellow.

The *head* is blunt, oval, small, and scaly. The *mouth* is low, and extends horizontally backward. The *jaws* are nearly of an equal length. The bones of the upper *lip* end blunt, as if cut transversely. The *tongue* is blunt, and tied down. In each jaw are crowded many *teeth*, with a single row of remote large ones intermixed. There are none on the *palate*. Both apertures of each *nostril* are circular; the foremost is surrounded by a margin. The *cheek-bones* are indented on the lower edge. The anterior plate of the *gill-covers* is also indented, with the four lower toothlets larger than the others. The hinder plate of the gill-covers terminates in a long strong spine. The *membranes* of the gill-covers are exposed, and each contains six rays.

The *back* has a furrow for receiving the fin. The *scales* are small, and finely indented on the edge, and adhere firmly. On each side of the nape is a scaly space, the margin of which, behind, is indented. Above each pectoral fin is a scale-like *appendant*, indented on its hinder edge.

The *back fins* are united but distinguishable, the edge of each forming an arch. The foremost contains ten strong prickles;

the hinder one prickle and ten soft rays, of which the last is divided to the root into two. The *pectoral* fins are small and acuminated above, and each contains thirteen rays. The *ventral* fins are large and remote, and each contains one prickle and five soft rays. The edge of the *vent* fin is rounded. It contains eleven rays, of which the three first are strong prickles, the others are divided, and the last is bifid to the root. The *tail* fin is forked, with blunt ends, and contains eighteen distinct rays, besides some short compacted ones at each side.

5th Species.—Coius gudgutia.

A *coïus* with the tail fin terminating in a straight line; with the united back fins containing twelve prickles and fourteen soft rays; with three prickles and seven soft rays in the fin behind the vent; and with the first branch of the second ray in each ventral fin lengthened in form of a bristle.

This fish is found in the mouths of the Ganges, and grows to about six inches in length. It is rather wide in proportion to its *length*, and much compressed, and its *colour* above is greenish, with faint longitudinal stripes, while below it is silvery. The back fins have faint black spots. The eyes are like silver.

The *head* is oval, bluntish; of a moderate size, and rough. The *mouth*, when shut, extends directly back. The under *jaw* is the shortest. The bones of the upper *lip* are small, and end blunt as if cut across. The *teeth* in both jaws are slender and crowded. The *tongue* is sharp, fleshy, and dotted. The *palate* is smooth. The apertures of the *nostrils* are oblong. The *eyes* are rather large, with the shortest axis across the length of the fish, while that of the pupil is in the contrary direction. The hinder edge of the *cheek-bones* is indented. Each *gill-cover* is rounded behind, covered with scales, and consists of two smooth edged plates. The *membrane* attached is not entirely hid, and contains seven rays.

The *back* is arched and rounded on the edge. The *belly* is flat. The *scales* are large, adhere firmly, and are indented on the edge. Above each ventral fin is a short scale-like *appendant.*

The first fin of the *back* slopes gradually behind from the point of the third ray, the two first being short, and contains eleven strong prickles. The edge of the second fin forms an arch. It contains one short prickle and fourteen soft rays, of which the first is undivided, and the last split into two to the bottom. The *pectoral* fins are of moderate size and sharp above, and each contains *seventeen* rays, of which the first, second, and seventeenth, are undivided. The prickle of each *ventral* fin is slender. The fin behind the *vent* slopes towards the tail; its foremost prickle is small, and its last soft ray is very deeply divided into two. The *tail fin* contains seventeen distinct rays, besides some short compacted ones at each side.

6th Species.—Coius polota. Plate XXXVIII. Fig. 31.

A *coïus* with the fin of the tail fan-shaped; with twelve prickles and an equal number of soft rays in the back fins; with three prickles and nine soft rays in the fin behind the vent; and with the second ray of each ventral fin elongated into two bristles.

This fish is found in the estuaries of the Ganges, but is not very common, nor is it considered good for eating. It is very tenacious of life.

The *form* is short, much compressed, and high-backed. The whole is scaly and rough to the feel. When of the ordinary size of four or five inches in length, it is of a very pale-green *colour* above and white below, is marked longitudinally with many fine waved shining lines, and transversely by eight broad black belts, of which the foremost forms an angle at the eye. When it is found of a large size, that is, a foot in length, the transverse belts are faintly marked, the longitudinal lines disappear, and each scale on the sides is marked with an oblong vertical white spot.

The *head* descends with a concavity above the eyes, and is narrow and scaly. The *mouth* is low, passes back horizontally, and is of moderate size. The under *jaw* is rather the longest. In each jaw is a row of sharp *teeth*, of unequal sizes. The *tongue* is sharp, free, and rough, and the *palate* is smooth. The hinder aperture of each *nostril* is oblong, the foremost is circular. The *eyes* are large and protuberant. The first plate of the *gill-covers* is indented on the edge, the hinder one ends in a point. The *membrane* of each gill-cover is half exposed, and contains six rays.

The *back* is much arched, and is furrowed for receiving the prickles of the fin. The *belly* is straight. The *scales* are large, adhere firmly, and are indented on the edge. The *shoulder* bones are indented.

The *back fins* are united, but distinguishable, each forming an arch, of which the first, as in the third and fourth species, is edged like a saw, and contains eleven prickles; the second arch contains one prickle and twelve soft divided rays. The *pectoral* fins are short and rounded, each having nineteen rays. The *ventral* fins are approximated with a strong prickle. The edge of the fin behind the *vent* forms an arch, and its prickles are strong. The *tail* fin contains seventeen distinct rays, besides some very short ones.

7th Species.—COIUS NANDUS. Plate XXX. Fig. 32.

A *coius* with the fin of the tail fan-shaped; with thirteen prickles and twelve soft rays in the united fins of the back; with three prickles and eight soft rays in that behind the vent; and with the body clouded with irregular dark marks.

In the ponds of the Gangetic provinces this fish is very common, seldom exceeds six inches in length, is very tenacious of life, and is considered by the natives as good for eating.

In proportion to its thickness, this fish is of a long form, and it is high-backed, and much compressed at the sides. The general *colour* is pale-green, with a silvery gloss, on which are scattered many large irregular marks, of a dark colour, inclining to olive. All the fins, except the pectorals, are spotted with the same colour.

The *head* is blunt, descends in a straight line, and is nearly oval, and large, but narrow. It is covered with scales to the lips, and between the eyes there runs a furrow. The *mouth* is low, and descends backward parallel to the lower *jaw*, which is rather the longest. The *lips* are thinner than in any species of this genus that I have seen. The bones of the upper one terminate very bluntly, as if cut across. On the jaws, palate, and tongue, are numerous small *teeth* crowded together; and intermixed with these, in each jaw, are several sharp teeth of a larger size. The *tongue* is sharp and free, and marked by two longitudinal ridges. On each side of the *palate*, at its fore part, there is a bone. Each *gill-cover* has four bony plates, of which the foremost is indented on the edge, and the uppermost ends in two sharp approximated points. The *membranes* of the gill-covers are partly exposed, and each contains seven rays. The bones of the *gills* are marked by a row of tubercles, of which the upper are covered with teeth.

The *belly* is straight, the *back* high. Where the tail becomes small, the *lateral line* is broken into two portions. The *scales* are finely indented on the edges; and on those of the sides each has on its centre a longitudinal mark.

The fins of the *back* are united, but readily distinguishable, the edge of each forming an arch. The first contains twelve strong prickles, and is indented like a saw on the edge; the second contains one prickle and twelve soft rays, of which the first is undivided, and the last split to the root. The *pectoral* fins are short and rounded, and each contains fifteen rays. The *ventral* fins are without bristles, and are short, and approximated to each other. The prickles of the fin behind the *vent* are strong and short. The *tail* fin contains in all sixteen rays, of which two on each side are undivided.

8th Species.—Coius cobojius. Plate XIII. Fig. 33.

A *coius* with the fin of the tail nearly rounded; with seventeen prickles and eight soft rays in the united fins of the back; with ten prickles and ten soft rays in the fin behind the vent; and with faint black belts transversing the sides.

This fish is the *Lutianus scandens* of La Cepede, (*Hist. des Poissons*, Tome IV. p. 242,) who seems to have borrowed his account from that given of the *Perca scandens* by Captain Daldorf, (Linnæan Transactions, Vol. III. p. 62,) a Danish, and not an English officer, as La Cepede supposes. He was a worthy man, very incapable, I believe, of advancing any thing that he thought untrue, and able enough to describe what he saw; but he did not possess those faculties which enable a man to reason on what he observes, or to take into account the collateral circumstances that might lead to an explanation of any thing extraordinary. Accordingly, he gravely assures us, that he saw one of these fishes climbing a palm-tree, and advancing up by the same means that are employed in moving along the ground. The French naturalist, to my utter astonishment, seems to think the circumstance merely curious; nor does he appear to have any doubt concerning the fish possessing the power of climbing trees. Nature, it is true, often gives faculties to animals, by which they are led to actions highly dangerous and pernicious to themselves; and moralists, and other dealers in trash, have therefore often declaimed against the ways of Providence, without reflecting, that these same faculties enable animals to procure the highest enjoyments of which their natures are capable. To what enjoyment this dangerous faculty of climbing trees could lead a wretched fish, I am totally at a loss to imagine, and I therefore believe that Daldorf was mistaken; but to what circumstance, neglected to be noticed in his narrative, the error should be attributed, I cannot take upon myself to say. The palm, as is often the case with those of its species, (Borassus flabelliformus,) may have been growing with its lower parts nearly horizontal, and the fish may have then moved along it, as well as on the land: or the palm may have been covered with the knobs, often left by the cultivators when they remove the branches, (stipites,) and the fish may have been left among these knobs by some bird, and might, no doubt, have continued wriggling among them.

Of all that I know, the *cobojius* is the fish most tenacious of life in the air; and I have known boatmen to keep them for five or six days in an earthen pot without water, and daily to use what they wanted, finding the fish as lively and fresh as when caught. In fact, the Calcutta market is chiefly supplied from extensive marshes in the Yasor district, and about one hundred and fifty miles distant. From thence boat loads are brought, and kept alive without water until sold.

This is one of the fishes supposed to fall with rain from the heavens; and I endeavour to explain the circumstances giving rise to this opinion, in the same manner as the similar supposition respecting the *Ophiocephalus gachua*, (page 68,) both fishes living in exactly similar situations, and both being possessed of great tenacity of life in the air, with a considerable facility of progressive motion on land.

The *cobojius*, which by the natives is considered as the prototype of their genus *coi*, from whence the name Coius is derived, seldom exceeds six inches in length, and is found every where in the marshes, ponds, and ditches of India. Although it abounds in sharp bones, the natives of Calcutta especially use it much in diet, the women thinking that it increases their milk, and the men imagining that it possesses great powers to invigorate.

The *form* of this fish is longer in proportion to its thickness than that of any *coius* which I have seen, and is less compressed at the sides, although this circumstance is abundantly evident. It is covered with rough scales, and above is of a dusky-green *colour*, and below of a pale yellow, with several transverse blackish bars on each side. The throat is tinged with blue. Sometimes, but not always, there is an irregular spot at the end of the gill-covers, and another at the end of the tail. Except the first fin of the back, the others have a reddish tinge. In dirty marshes the whole fish becomes nearly black.

The *head* is oval, blunt, little compressed, wider than the body, but of moderate size, and is covered entirely with scales. Before and beneath each eye is a *bony process*, indented on the lower edge, and ending before in a sharp point, which the animal can at pleasure turn out, and which is an organ of progressive motion, when it is on land. The *mouth* is near the middle of the head, passing backward with a curve. The *jaws* terminate the head, and are nearly equal in length. The outer *teeth* in each jaw are disposed in a row, remote from each other, and are large and conical. Within these are crowded many small ones. The *tongue* is smooth, blunt, and free. The *palate* is smooth. The foremost aperture of each *nostril* is tubular. The *eyes* are placed very far forward, and are reddish. Each *gill-cover* consists of three plates, the two hindmost of which are indented with numerous strong parallel spines; and, while the fish is on land, are powerful organs of locomotion. The *membranes* of the gill-covers are mostly exposed, and are so thick, and covered with scales, as to conceal the rays, if they contain any. The bones of the *gills* are provided on the inside with a row of tubercles.

The *back* and *belly* are nearly equally prominent, the former having a furrow for the first fin, as has the *tail* for the fin behind the vent. The *lateral lines* are broken into two portions, opposite to the hinder prickles of the back and vent fins. The *vent* is rather behind, but very near the middle. The *scales* are large, dotted, and indented round the edge.

The *back* fins, although fully united, are distinguishable. The foremost has throughout nearly the same height, and an edge indented like a saw, and contains all the prickles which are strong. The second ends somewhat sharply behind. The *pectoral* fins are blunt, but longest above, and each contains sixteen branched rays. The *ventral* fins have no bristles, and are short. The fin behind the *vent* consists of two united; the foremost having an arched indented edge, and being supported by prickles; while the hinder is higher, ends in a sharpish point, and contains the soft rays. The *tail* fin is nearly round, approaching a little to the fan-shape, and contains about fourteen rays.

9th Species.—Coius chatareus. Plate XIV. Fig. 34.

A *coius* with the tail fin ending in a concave edge; with one back fin containing five prickles and twelve soft rays; with three prickles and sixteen soft rays in the fin behind the vent;

and with a silvery body, having on the back large black spots.

This fish, which is rather anomalous, is found in the mouths of the Ganges, grows to about a foot in length, and is considered as very poor eating. Contrary to what happens in the *form* of all the other species, the greatest width is behind the middle, and it is more prominent below than above. It is much compressed on the sides. The greater part is of a bright silvery *colour*, with a little green on the back, where, on each side, there are five or six large oblong irregular black spots. The back fin approaches to an olive colour, and at the hinder end has a large black spot. The fin behind the vent toward the edge is blackish. The eyes are silver coloured.

The *head* is oval, sharpish, and small; above smooth, and below rough. The *cheek-bones* are indented on the lower edge. The *mouth* is very large, and, when shut, descends obliquely backward. The under *jaw* is the longest, and is covered with scales, as are also the bone of the upper *lip*, which are narrow. The *teeth* in both jaws are sharp, slender, very short, and crowded in great numbers. The *tongue* is sharpish, osseous, rough, and free. The *palate* is covered in part by three rough bones, of which the two at the sides are very large and parallel. There are two round bones at the entrance into the throat. The foremost aperture of each *nostril* is tubular, the hinder one circular. The *eyes* are large and circular, with the greatest axis of the pupil in a vertical position. The *gill-covers* are rounded behind, each consisting of two smooth edged plates. Their *membranes* are partly exposed to view, each containing seven rays.

The *back* descends very gently. The *belly* is very prominent, and its ridge is sharpish. The *vent* is near the middle. The *scales* on the upper parts are large and smooth edged; on the lower parts they are smaller and indented. Above each ventral fin is a very large scale-like *appendant* ending in a sharp point.

There is only one *dorsal fin* placed far back, and covered half way up with scales. Its edge forms an arch deeply indented on the fore part, the membranes intervening between the prickles extending to the point of the prickle before, and only to about the middle of that behind. The prickles are strong, and the last of the soft rays is very deeply divided. The *pectoral* fins are very short and sharp above, each containing twelve rays, of which the first and last are undivided. The *ventral* fins are small, and placed so near each other as almost to unite. They are indented, the edge of each membrane forming a concavity. The prickle in each is strong, and almost as long as the divided rays. The edge of the fin behind the *vent*, which is covered with a scaly skin, forms an arch indented at the prickles, which are strong. The *tail* fin is blunt at both extremities, with a concave edge between, and contains seventeen distinct rays, besides some short compacted ones at each side.

VIII. Genus.—CHANDA.

Fishes of the fourth order, with sharp teeth scattered in both jaws, with some of the bones of the head distinctly indented on the edge; with prickles in some of the fins; and with bodies more or less diaphanous, and much compressed, the sides being nearly perpendicular.

The fishes of this genus are all very small and of little value, although in many places abundant, and used in considerable quantities; but as food they are insipid, and filled with small bones, for which defects their size does not compensate. They form a very natural arrangement, and the first has the strongest affinity to the *Zeus insidiator*, so that all of them, no doubt, belong to the same genus with that fish, although I cannot help thinking, that, to include them in the same genus with the *Zeus ciliaris* and *Zeus faber*, is an unnatural arrangement. Owing to the smallness of several species, the teeth are minute, and in some are more distinguished by a feel of roughness than by being visible to the naked eye; but, in proportion to their size, they are thick at the base, and end in a sharp point.

Besides the circumstances mentioned above, and in the generic character, all the species, which I have described, have the following in common.

The *head* is devoid of scales, except a very few on its hinder part, which may be observed in some species. The *jaws* are protruded in opening the mouth, and, in most species, when thus opened, form a kind of *trunk*. Each of the membranous *covers* of the gills contains six rays.

The *back* slopes from the fin to the nape, where there is generally a concavity, the slope of the head being less than that of the back. The edges of the scales, where there are any, are not indented.

Where there are two distinguishable *back fins*, the first contains seven prickles, and the last one. Each of the *ventral* fins contains six rays, of which the first is a prickle. The fin behind the *vent* always contains three prickles, besides soft rays.

As in the genera already described there are, as it were, certain intermediate species, so in this the two first, which I have described, together with the *Zeus insidiator*, have but little of the transparency, which forms part of the generic character, but the resemblance to those which follow is so strong, that all natives give them a common appellation.

1st Species.—CHANDA? SETIFER.

A *chanda?* with one back fin containing ten prickles, and nine soft rays; with seven soft rays in the fin behind the vent; with the first branch of the second ray in each ventral fin lengthened out in form of a bristle; and with scales on the body, which is almost opaque.

This species is ill defined, and might, perhaps, be placed as a *coius.*

This fish grows to about four inches in *length*, and is found in the estuaries of the Ganges. It is narrower than the *ruconius*, of a green *colour* above, and silvery below, and has no spots.

The *head* is oval, sharp, small, and behind covered with scales. The under *jaw* is rather the shortest. There is some appearance of *lips*. There is a roughness on both jaws in place of *teeth*. Each *nostril* has two circular apertures. The *eyes* are lateral, large, and somewhat oblong, with circular pupils, and silvery irides. The *bones* under the nose, as well as those of the cheeks, are indented on their lower edges. Each *gill-cover* is rounded behind, and consists of two plates, that have smooth edges. Their *membranes* are half uncovered.

The *back* is prominent, and has a sharp ridge. It is provided with a *furrow* for lodging the fin of the back. The *belly* is straight. The *lateral lines* are arched parallel to the back. The *vent* is behind the middle. The *scales* are large, and adhere firmly. Above each ventral fin is a scale-like appendant, of moderate size.

The edge of the *back* fin forms an arch, with an indent at the end of the part supported by prickles, the first of which is very short. There are nine soft rays, and they are branched at the top. The *pectoral* fins are low, very long, and acuminated above; each contains sixteen rays, of which the two first and last are undivided and short. The *ventral* fins are short and near each other: their prickle is strong. The fin behind the *vent* is high before, and toward the end gradually becomes lower: besides the prickles, of which the first is very short, it contains seven branched rays, of which the last is split to the root. The *tail* fin is deeply divided, and contains seventeen distinct rays, besides some short compacted ones.

2d Species.—CHANDA RUCONIUS. Plate XII. Fig. 35.

A *chanda* with one dorsal fin, containing eight prickles and sixteen soft rays; with fourteen soft rays in the fin behind the vent; and with the body covered with scales slightly diaphanous, and of a silver colour, having transverse dotted black bars across the back.

This may perhaps be the *Zeus kleinii* of Walbaum in his edition of Artedi, a fish which I cannot trace in Bloch or La Cepede. It has also a strong resemblance to the *Komah karah* of Russell, (Indian Fishes, Vol. I. No. 63;) but the latter has more of a rhomboidal form, yellow bars on the back, and larger teeth; nor does Dr Russell mention, that any of the bones of the head are indented on the edge.

The *Ruconius* or *Ruconi chanda* of the natives is found in the estuaries of the Ganges, and is very nearly allied to the *Zeus insidiator*, which the natives of Bengal call *Ectuntia chanda.* It is a very beautiful fish, seldom exceeding three inches in *length.* It is oval, but rather sharp at both ends, and is short and narrow in proportion to its height. The whole, except the fins, which are clear and unspotted, is of a bright silver *colour*, with many transverse black bars, formed of numerous dots, and descending to the middle of each side. Although the bright colour of the scales conceals the ribs, the muscles are visible.

The *head* is small, oval, naked, and above each eye is armed with two small prickles. The lower edge of the *cheek-bone* is indented. The *mouth* is nearly terminal, and when opened, the *jaws* form a conical trunk half as long as the head. The *nostrils* are near the eyes, and each has only one aperture. The *eyes* are large, and stand high. Each *gill-cover* consists of four plates free from scales, and rounded behind, the uppermost of the two before having its edge indented. The *membranes* of the gill-covers are entirely concealed.

A sharp-pointed naked bone runs down the slope of the *back*. Both back and *belly* are very prominent. The *vent* is before the middle, and the *tail* above and below is indented at each ray of the fins. The *lateral line* is high, and there is an appearance of other lateral lines formed by the union of the muscles which act in contrary directions. The *scales* are minute and smooth, and adhere firmly. Above each ventral fin is a long scale-like *appendant.*

On the *back* there is only one fin, which grows gradually lower and lower from the second ray, which is about twice as long as the last; the first is short; the prickles are strong, and the fore part of the third and fourth is indented. The *pectoral* fins are low, short, and sharp above; each contains sixteen rays, of which the first is undivided. The *ventral* fins are close to the vent, and short. The fin behind the *vent* is shaped like that of the back, and the third prickle is indented on the fore side. The fin of the *tail* is divided into two bluntish lobes, and contains seventeen distinct rays, besides some short compacted ones on each side.

3d Species.—CHANDA NALUA. Plate VI. Fig. 36.

A *chanda* with the back fins united, the second containing eleven soft rays; with ten soft rays in the fin behind the vent;

with the head naked; and with the body covered with large scales.

If this fish be different from the *Centropomus ambassis* of La Cepede, (*Hist. des Poissons*, Tome IV. page 252,) it is certainly a very similar species.

The *Nalua* is found in the fresh water rivers of the lower parts of Bengal, and is high in proportion to its *length*, although not so much in degree as the second species is; and its back projects more than the belly, although both are prominent. Above it is of a pale green *colour*, and below is diaphanous with a purple gloss, so that the back-bone is discernible. The cavity of the belly is lined with a silvery membrane, visible through the sides; and the same colour shines on the head and eyes. The first fin of the back in its fore part is dotted with black, and over the back-bone are five or six black spots disposed in a row.

The *head* is oval and rather narrow, and between the eyes there runs down a ridge, which, toward the mouth, divides into two. The *mouth*, when shut, descends almost vertically, and is large. The *jaws* are protruded in opening, but scarcely form a trunk; the lower is the longest. The *teeth*, though small, are distinctly visible. The bones of the upper *lip* are pretty large, and widen below. The *tongue* is sharp, and moves freely. Each *nostril* has two apertures, the hinder one oblong. The *eyes* are high, large, and circular. The bone under the nose, and that of the *cheek*, is indented on the edges. The *gill-covers* are devoid of scales, and each consists of four plates, of which the two foremost are indented on the edge. They conceal their *membranes*.

The *back* is arched, and forms a sharp ridge. The *lateral lines* are high, and arched parallel to the back. The *vent* is nearly in the middle. The *scales* are large, and adhere firmly; those on the back are dotted round the edges; while, on the belly, owing to their pellucidity, they are scarcely visible.

The two fins of the *back*, although in some degree united, are very readily distinguishable. The first is rather the highest, and is supported by seven strong prickles rather longer than the intervening membrane. The first prickle is very short, the second is the longest, and the subsequent ones gradually diminish, so that the last is not above half the length of the prickle of the second fin, of which the soft rays are branched, and grow shorter and shorter to its end, which is rather sharp. The *pectoral* fins are rather small, situated low, and sharp above: each contains thirteen rays. The *ventral* fins are short. Their prickle is strong, and the membranes intervening between the rays do not reach to their ends. The first prickle of the fin behind the *vent* is short, the second is the longest, and the length of the other rays gradually diminishes, as in the second fin of the back. The *tail* fin contains fifteen distinct rays, besides some short compacted ones on each side.

4th Species.—Chanda nama.

A *chanda* with two back fins united, the second containing sixteen soft rays; with fifteen soft rays in the fin of the vent; and with the body long, and devoid of scales.

This fish is common in ponds throughout Bengal, and seldom exceeds three or four inches in *length*. The back and belly form arches nearly equal, and slope gradually to the mouth, and end of the tail. In respect to *colour*, the upper part of the head, the back, belly, and tail, are diaphanous, so as to discover the ribs and back-bone, a silver line running along the latter, and the former being seen over a silvery membrane that lines the visceral cavity. The lower parts of the head, especially the gill-covers, are opaque and silvery. Over the back and its fins, together with those of the tail and vent, are scattered black dots, which, on each shoulder, are collected into an oblong mark, and into another on the summit of the back fin.

The *head* is oval, and rather small. The *mouth*, when shut, descends very obliquely, and is large. When open, the *jaws* protrude only a little; the under one is by much the longer. The bone under the nose is not indented on the edge; but the *cheek-bone* has some teeth at its hinder angle. The *lips* are somewhat fleshy, with scarcely any bones in the upper. In both jaws large conical *teeth* are scattered at a distance from each other. Each *nostril* has one aperture in the middle between the eye and jaw. The *eyes* are large, high, and far forward. Each *gill-cover* consists of four plates, of which the foremost has an indented edge. The *membrane* covering the gills is exposed to view.

The ridge of the *back* is sharp. The *tail*, at the fins of the back and vent, is indented on the edges. There is no *lateral line*. The *vent* is near the middle. There are two *back* fins united, but clearly distinguishable. The prickles of the first, in number seven, project a little beyond the intermediate membranes. The first is very short; the second is the longest; and the others diminish suddenly in length to the last. The second fin contains one prickle, and its soft rays are branched, gradually shortening from the prickle, which is the longest, and three times the length of the last ray of the first fin. The *pectoral* fins are low, short, and sharp above: each contains eleven rays, of which the first and second are undivided, and the former is very short. The *ventral* fins are short, and have even edges. The prickles of the fin behind the *vent* become gradually longer and longer to the third, from whence the divided rays gradually shorten to the last. The lobes of the *tail* fin are sharpish, and contain seventeen distinct rays, besides several short compacted ones on each side.

5th Species.—CHANDA PHULA.

A *chanda* with two back fins united, the second containing fourteen soft rays; with thirteen soft rays in the fin behind the vent; and with the body long, transparent, and devoid of scales.

This fish, which is found in the ponds and rivers of the north-eastern parts of Bengal, strongly resembles the *Nama* last described, but seldom exceeds two inches in length. The *colours*, however, are exactly the same, even to the small collection of dots on each shoulder; nor, indeed, except in the number of rays which support the fins, is there any considerable difference between the two species. I shall only, therefore, state these little differences.

The *head* is half oval, being nearly horizontal above. When open, the *jaws* protrude considerably. The *nostrils* are near the eyes. The edges of the *cheek-bones* are not indented. The rays of the *membranes* covering the gills are so minute, that I am not certain as to their number, although I think that I could count six.

The *vent* is before the middle.

The lower lobe of the *tail* fin is longer than the upper.

6th Species.—CHANDA BOGODA.

A *chanda* with two back fins united, the second containing sixteen branched rays; with seventeen soft rays in the vent fin; and with a long transparent body devoid of scales.

This species, which is also found in the north-eastern parts of Bengal, and is no longer than the *phula*, differs in nothing remarkable from the two former, except in the number of soft rays contained in the back and vent fins, and in that contained in the pectorals, each of which has twelve.

7th Species.—CHANDA BACULIS.

A *chanda* with two fins of the back united, the second containing thirteen soft rays; with an equal number in the fin behind the vent; with the body short, transparent, and devoid of scales; and with a yellow mark on the nape.

This species, which is also found in the north-eastern parts of Bengal, connects strongly the three preceding fishes with the two which follow, resembling the former in colour, and the latter in shape. It seldom exceeds an inch and a half in *length*, is much compressed, and, in proportion to its length, is higher than the three preceding species, being nearly oval. In regard to *colour*, it is so diaphanous that the muscles are visible, forming transverse bars on the sides; but, in other respects, differs little from the three preceding species, only there are no dots on the body, except in a large yellow spot on the nape of the neck.

The *head* is of moderate size, narrower than the body, oval, and sharp. The *mouth* is small and high, and when it opens, forms a short trunk, the upper *jaw* being the shorter. The *teeth* are minute, the *nostrils* have double apertures, and the *eyes* are high and circular. The *bone* under the nose has its lower edge indented, as is also the case with the anterior plates of the *gill-covers.*

The ridge of the *back* is sharp, and the *tail* has a sharp ridge at the insertions of the back and vent fins. The *vent* is before the middle.

The two *back* fins are united, but most clearly distinguishable. The first slopes gradually from the second to the seventh and last prickle, the first being very short. The first ray of the second fin is a long prickle. Each *pectoral* fin has about ten rays. That of the *tail* has seventeen rays, besides some short compacted ones on each side.

8th Species.—Chanda ranga. Plate XVI. Fig. 38.

A *chanda* with two back fins united, the first indented on the edge, and the second containing one prickle and twelve soft rays; with fifteen soft rays in the fin behind the vent; and with a short body covered with scales.

This is a beautiful little fish, about an inch and a half long, and is found in the fresh waters of all parts of the Gangetic provinces. It is of a rhomboidal *form*, and, with regard to *colour*, has a bright green and silver gloss. The whole, except the gill-covers, which are silvered, is diaphanous, showing the muscles, five ribs on each side the back-bone, a silver stripe accompanying the latter, and a silvery membrane that lines the cavity of the abdomen. The upper parts are dotted with black.

The *head* is oval and small. The *mouth* is small, and, when shut, descends obliquely. The *jaws*, when protruded, form a short trunk; the under one is the longest: neither has *lips*. The lower edge of the bones under the *nose* is indented, as is that of the *cheek-bones*. In both jaws there are minute *teeth*. There is scarcely any *tongue*. The fore part of the *palate* is tuberculated. Each *nostril* has two circular apertures in the middle, between the eyes and nose. The *eyes* are large and high. Each *gill-cover* has four plates, of which the foremost has indented edges. The *membranes* covering the gills are exposed.

The *back* has a sharp ridge; both it and the *belly* are equally prominent, and slope gradually to the head and tail, but more suddenly to the former. The *lateral line* is more distinct than in the four preceding species, and forms a small arch above the ventral fins, behind which it descends to the middle of the fish. The *scales* on the body and fins are minute, and so transparent, that they can scarcely be seen without a magnifying glass: they are marked with concentric lines.

There are two *back* fins united, but most readily distinguishable. The prickles in the first, seven in number, are longer than the intermediate membranes, and gradually become shorter and shorter from the second, which is the longest, and much longer than the first. The edge of the second fin forms an arch, its prickle is twice as long as the last of the foremost fin, and the soft rays are branched. The *pectoral* fins are short, sharp above, and low on the sides. The *ventral* fins are short. The edge of the fin behind the *vent* is arched, but the membrane between the prickles is short. The fin of the *tail* contains eighteen distinct rays, besides some small compacted ones at each side.

9th Species.—Chanda lala. Plate XXI. Fig. 39.

A *chanda* with two back fins united, the first being even on the edge, and the second containing from twelve to fourteen soft rays; with fifteen soft rays in the fin behind the vent; and with a short scaly body.

This fish is found along with the last described, and is still more beautiful: indeed, nothing can exceed the delicacy of its tints; but it differs more in general appearance than in any part of its general structure from which a satisfactory specific character could be drawn; especially as, in young individuals, the *colours* of the two species are faint and ill defined. In the adult *lala* the gloss is golden, and the dots are orange. The sides have several transverse bars, alternately of dusky and of gold, with a green gloss.

The most remarkable difference in the structure is, that the rays of the back and vent fins in this species are longer in proportion than those of the *ranga,* and slope more toward the tail, while the intervening membranes in the first fin of the back have an even edge, equal in length to the rays.

IX. Genus—TRICHOPODUS.

Fishes of the fourth order, with only one bristle, in place of each ventral fin.

The six following fishes are evidently of the same genus with the *Trichopodus trichopterus* of La Cepede, (*Hist. des Poissons,* Tome III. p. 129,) from which they differ in having the end of the tail fin rounded; but form a very natural well defined genus, which agrees with the last described in every thing, except the form of the ventral fins. These in reality are placed before the pectorals; so that, in strictness, this genus should be placed among the *jugular* fishes next to the *Oligopodus* of La Cepede, (*Hist. des Poissons,* Tome II. p. 311.) From its strong affinity, however, with the three preceding genera, and with that which follows, I have imitated La Cepede in placing it in the *thoracic* order.

The species of *trichopodus,* which I have seen, are all very beautiful, and are common in the ponds, marshes, and ditches of the Gangetic provinces; but they are all small, are no where found in great quantity; and, therefore, although eaten by all ranks using fish, and thought palatable, they are not of consequence. They all possess the following characters.

They are of an oblong *form,* but have a considerable height, while their sides are much compressed, and nearly vertical. They are rough to the feel, opaque, and variegated with bright and agreeable *colours.*

Their *heads* are oval, rather small, and covered with scales even on the throat, while some of the bones have indented edges. The *mouth* is small and high, and, when shut, has a great descent: when open, the *jaws* protrude far, so as to form a kind of trunk. The *teeth* are either altogether wanting or very minute, so as, in general, to be a mere roughness, distinguishable more by the feel than by sight. Each *nostril* has a double aperture. The *eyes* are high on the sides of the head, and generally far forward. The *gill-covers* are scaly, move freely, and leave an ample *opening* at each side.

The *back* and *belly* are both nearly equally and regularly arched, nor, in general, is there any sudden contraction from the mouth to the tail fin; but the greatest width is before the middle of the fish. The *vent* is far before the middle. The edges of the *scales* are finely indented.

The *back* fin extends from the apertures of the gills to near the fin of the tail, and rises gradually from the first prickle to near the end. It consists of two parts, distinguished by the nature of their rays, but not separated by any division or interruption. The first part, which is the longest, is supported by prickles, and is indented on the edge like a saw: the second part, which is the highest, is supported by soft rays. The *pectoral* fins are fan-shaped, and short. The *ventral* fins have no membrane, are placed before the pectorals, and each consists of one soft ray, shaped like a bristle, and very long, extending near, at least, to the commencement of the tail fin. The fin behind the *vent* resembles in structure that of the back, and extends almost the whole length of the tail. The fin of the *tail* is fan-shaped; but, in some species, has more or less of a concavity on its hinder edge.

The Bengalese give to all the species the generic name *Colisa, (Kholisha;)* but the first, being considered as the prototype of the genus, receives this name without addition, while each of the others receives a specific appellation prefixed to the generic name.

1st Species.—TRICHOPODUS COLISA. Plate XV. Fig. 40.

A *trichopodus* with the fin of the tail fan-shaped; with transverse bars on the sides of green and pale yellow; and with the back fin rounded behind.

This species is sometimes found five inches long. Above it is of a fine green *colour*, below whitish, with many pale brass-coloured bars descending obliquely back along the sides. The vent fin is edged with red, and it is variegated with black, green, and white. The back and tail fins are spotted with black, and on the hinder part of the former are an intermixture of red spots. On each gill-cover there is a green spot. The eyes are red.

The *head* is blunt, and under each eye is armed with a sharp-pointed *bone* indented on the lower edge, the fore end of which the animal can at pleasure turn outward. The lower edge of the *cheek-bone* is indented. The *teeth* are very minute. The *tongue* is scarcely observable. The *palate* is smooth. The *nostrils* are circular. The *eyes* are of a moderate size, with circular pupils. Each *gill-cover* consists of two plates, ends in a point behind, and is indented on the lower edge. The *membrane* of the gill-covers is exposed on the throat. I can observe in it no rays, owing, perhaps, to its thickness. The bones of the *gills* are smooth on the inside.

The *back* and *tail* have a furrow for receiving the prickles of their fins. The *lateral line* is high and straight, and, opposite to the last prickle on the back, is broken into two parts. The *scales* are large, adhere firmly, are dotted before, and marked towards the ends with diverging rays.

The *back* fin contains sixteen strong prickles, and eleven

branching rays. The *pectoral* fins are near the middle height of the side, each containing ten rays. The *ventral* fins reach to the root of the tail fin. The fin behind the *vent* contains seventeen strong prickles and an equal number of branched rays. The tail fin has sixteen rays, with a very short one at each side.

2d Species.—Trichopodus bejeus.

A *trichopodus* with a notch in the edge of the tail fin; with many transverse bars on the sides of green and pale yellow; with the head armed beneath each eye by an indented sharp bone; and with the end of the back fin sharp.

I have not seen this fish longer than four inches. The upper parts are of a dusky green *colour*, and the lower are white, while the sides are transversely marked by the green and copper-coloured bars. The back fin is greenish, with black spots on its fore part, and red spots behind. The fin behind the vent is spotted like that of the back, and edged with red. The tail fin is spotted with black and red, and the eyes are silver-colour, stained with red. On the whole, it has the strongest resemblance to the first species.

The *head* is blunt, and armed with a *bone* below each eye, as in the first species. The edge of the *cheek-bone* is indented. In both *jaws* there are very minute, but observable *teeth*. The *eyes* are of moderate size. The *gill-covers* are rounded, and indented on the edges. Their *membranes* are concealed, each containing about three rays.

There are no proper *lateral lines;* but each scale having on its centre a long elevation, these give an appearance of several lateral lines.

The *back* fin terminates behind in a sharp point, and contains seventeen prickles and nine soft rays, of which the third is the longest. Each *pectoral* fin contains nine rays. The *ventral* fins reach to the end of the tail fin. The fin behind the *vent* ends behind in a sharp point, and contains sixteen prickles and seventeen soft rays, of which the seventh is the longest. The *tail* fin is shaped like a fan; but in the centre of its edge there is a notch. It contains sixteen long rays, and a short one at each side.

3d Species.—TRICHOPODUS COTRA.

A *trichopodus* with the fin of the tail fan-shaped, and with transverse bars on the sides of green and blue.

I have not seen this fish longer than two inches. Above it is of a green copper *colour:* beneath it changes with copper and silver glosses, and on the tail is added a gloss of purple. On the sides are many transverse blue bars. The back and tail fins have dusky spots; that of the vent has blue spots and a red edge.

The *head* is sharp, and, before the eyes, unarmed, and the edges of the *cheek-bones* are not indented. No *teeth* are observable. The *eyes* are rather small, but protuberant. The foremost plate of the *gill-covers* has an indented edge. Their *membranes* are concealed, with about four rays in each, but the rays are indistinctly seen.

At the *nape* there is a kind of concavity, the slope being there somewhat sudden. The *belly* terminates in a ridge. There are no *lateral lines.*

The *back* fin contains seventeen prickles and eight soft rays. Each of the *pectoral* fins has nine rays. The *ventral* fins extend to the last prickle of the back. The fin behind the *vent* contains nineteen prickles and sixteen soft rays. The *tail* fin contains seventeen rays.

4th Species.—Trichopodus lalius.

A *trichopodus* with the fin of the tail fan-shaped, and with transverse red bars on the sides.

This most beautiful fish is no larger than the third species, and is of a green *colour*, with many transverse red bars on the sides, and with a silver-coloured gloss on the breast and gill-covers. The hinder fins are spotted with red. The eyes are red with a dusky transverse bar.

The *head* is blunt and unarmed, and the edges of the *cheek-bones* are not indented. There are no *teeth*, except a roughness on the jaws. The *eyes* are of a moderate size. The edge of the foremost plate in each *gill-cover* is indented. The *membranes* are concealed, each containing about four rays.

There is no *lateral line*.

The *back* fin contains sixteen prickles and eight branched rays. The *pectoral* fins are low, and each contains nine rays. The *ventral* fins reach to near the end of the fin behind the *vent*, which contains eighteen prickles and fifteen branched rays. In the *tail* fin there are sixteen distinct rays, with some short compacted ones on each side.

5th Species.—Trichopodus sota.

A *trichopodus* with the edge of the tail fin slightly concave; and with the upper and hinder parts brownish red, and the fore and lower parts silver-coloured, with a blue gloss.

This fish seldom if ever exceeds an inch and a half in *length*. Its upper and hinder parts are of a dusky reddish *colour*, very faintly marked with longitudinal brownish lines. The lower and fore parts are silver-coloured, with a blue gloss. The body has many dots, but no considerable spot. The hinder part of the vent fin is edged with red.

The *head* is sharp and unarmed before the eyes. The edges of the *cheek-bones* are not indented. There are no *teeth.* The *eyes* are small, but protuberant. The foremost plate of each *gill-cover* is indented on the lower edge. The *membranes* covering the gills are concealed, and each contains about four rays.

There are no *lateral lines.*

The fin of the *back* contains eighteen prickles and eight branched rays. The *pectoral* fins are low, and each contains nine rays. The *ventral* fins extend to the end of the fin behind the *vent,* which contains nineteen prickles and eleven soft rays. The fin of the *tail* contains about sixteen rays, and its hinder edge forms a slight concavity.

6th Species.—TRICHOPODUS CHUNA.

A *trichopodus* with the edge of the tail fin concave behind; and with a longitudinal black stripe, glossed with gold on each side.

This fish is about the *size* of the fifth species. It is above of a brownish-green *colour,* and below whitish, with a purple gloss on the tail; while, from the eye to the end of the tail, on each side, there runs a longitudinal stripe consisting of black dots, and shining with a golden gloss. The eyes above are scarlet. The fin behind the vent is edged with red.

The *head* is sharp and unarmed. The foremost plate of each *gill-cover* is indented on the edges.

The fin of the *back* contains seventeen prickles and seven soft rays. The *pectoral* fins contain each nine rays. The *ventral* fins are as long as the tail. The fin behind the *vent* contains nineteen prickles and eleven soft rays. The hinder edge of the *tail* fin ends in a slight concavity; it contains fifteen distinct rays, besides some short ones at the sides.

X. Genus.—Chætodon.

Fishes of the fourth order, with the edges of all the bones of the head smooth; with the sides much compressed, covered with scales, and high in proportion to the length of the body, and having in each jaw a single row of slender teeth parallel and close to each other.

1st Species.—Chætodon pairatalis. Plate XIV. Fig. 41.

A *chætodon* with the tail fin rounded at the end; with eleven prickles and sixteen soft rays in the back fin; with the body spotted; and with a bristle-like elongation from each ventral fin.

I should have supposed this fish to be the *Chætodon argus* of naturalists, were it not for the following circumstances: These authors take no noticè of the bristle-like production from the ventral fins. They represent the posterior edge of the tail fin as having the form of a crescent, with a slight convexity in the middle, so that it may almost be said to be trifid. Dr Russell (Indian Fishes, Vol. I. No. 78) represents the colour of the argus to be brownish, with a brighter gloss below, and a yellowish breast; and the spots on his fish are fewer in number, and smaller, than on ours. In the figures given by Bonnaterre, (*Encyclop. Method.*) and by Bloch, (*Ichth.* Planche CCIV. Fig. 1,) the spots resemble those described by Dr Russell, and in Bloch are made to extend over the fins of the back, and behind the vent, which, in our fish, have no spots. The figure of Bloch besides represents the gill-covers as devoid of scales.

The *Pairatalis* is found in the mouths of the Ganges, and, when newly caught, is a fish of great beauty, easy digestion, and excellent flavour; but, after death, it soon becomes soft and strong tasted. It grows to a foot in length, is very narrow in proportion to both height and length, and is somewhat of a quadrangular *form*, sloping very suddenly both before and behind, and its height being little short of its length. Its *colours* vary much in different individuals; but in all cases the sides, so far down as the pectorals and lower part of the tail fin, are covered with fine round black spots, larger than the spaces that intervene. The most common colour of the adult fish is blue, with a tinge of green on the back, and a silvery gloss on the sides, especially on their lower parts; and the hinder fins are yellow. In many individuals, however, the back is green, and the lower parts ash-coloured, changing to purple, with a gloss of silver. In young individuals, again, the whole is green, with a bright golden gloss. The eyes are black, with a narrow golden ring surrounding the pupil.

The *head* is small, oval, and blunt, with a convexity between the eyes. The *mouth* is small, and extends horizontally back from the extremity of the head. The *jaws* are nearly equal in length. The *lips* are fleshy, and without bones. The *palate* is smooth. The *nostrils* are in the middle, between the eyes and jaw, each containing two apertures, of which the foremost is circular and open, and the hindmost is oblong, and covered by a reflected membrane. The *eyes* are high, on the sides of the head, are of moderate size, and circular, with oval pupils, pointing forward. The *gill-covers* have minute scales, end in a point behind, and on each side leave a large *aperture*. Each consists of two plates, and exposes to full view its *membrane*, which contains four rays.

The fore part of the *back* descends very rapidly on the head. Behind it contains a furrow for receiving the prickles of the fin. The *belly* forms an arch less prominent than the back. The *tail*, both above and below, slopes very rapidly near the end, and before the slope has a furrow for receiving the prickles of the vent fin. The *vent* is behind the middle. The *lateral lines* are high, and run parallel to the back. The *scales* adhere firmly, and are small on the body, and minute on the head; but all have indented edges.

The *back* fin consists of two united, but distinguishable. The foremost contains ten strong prickles, and is indented like a saw on the edge. The prickles from the first, which is very short, increase rapidly in length to the fourth, which is the longest; and from thence they decrease gradually in length to the last, which is but little longer than the first. The posterior fin, covered half way with scales, occupies the slope on the tail, and with its ridge forms a vertical arch, its rays shortening very gradually towards the end of the tail: the first is a short prickle, which is not much more than half the length of the next, that is undivided; the remaining sixteen are branched. The *pectoral* fins are low, short, and blunt, but their upper rays are the longest. Each contains eighteen rays, of which the first, second, and eighteenth, are undivided. The *ventral* fins are near each other, each having six rays, of which the first is a very strong prickle, and the others are branched, the first branch of the second ray being lengthened out into a long awl-shaped flexible appendant. Behind the *vent* there are two fins slightly united into one. The edge of the foremost forms a concavity between every two prickles, and slopes gradually towards the end. It contains three strong prickles, of which the second is rather longer than the first. The hinder fin is covered half way with scales, and placed on the slope of the tail; and its rays decrease gradually backward from the second, which is the longest. The first is a short prickle; the second is soft, but undivided; and the other thirteen are branched. The *tail* fin is fan-shaped, and contains fourteen distinct rays, besides some short compacted ones at each side.

When employed in taking an account of Malabar, I had an

opportunity of describing another species of *Chætodon*, of which I shall here insert an account.

Chætodon caris.

A *chætodon* with a slight concavity in the posterior edge of the tail fin; with eighteen prickles and fourteen soft rays in the fin of the back; with the upper parts green and the under white; and with five broad yellow bars transversing each side.

This fish has a considerable affinity to the *Chætodon suratensis* of Bloch, (*Ichth.* Tome VII. p. 1, Planche CCXVII.) but that fish is purple and brown, where the *caris* is green and yellow. Besides, its tail fin is rounded at the end; its mouth is more prominent; and it has only thirteen rays in the back fin.

The *caris* is found in the clear rapid streams of Malabar, grows to about a span in length, and is tolerably good for eating. It is very much compressed, and in *form* resembles an elipsis. Besides the *colours* mentioned in the specific character, numerous small white spots are scattered on the body, and behind each pectoral fin there is a large black spot.

The *head* is scaly, oval, short, and rather wider than the body. The *mouth* projects very little, extends directly back, and is of moderate size. The *jaws* are nearly equal in length, when open are protruded, and the upper *lip* has no bones. The *nostrils* are in the middle, between the eyes and jaw, and each has two small apertures. The *eyes* are high, and rather far back, protuberant, of moderate size, and round, with oblong pupils. The *gill-covers* have scales, and are arched behind, each consisting of three plates, and exposing half its *membrane*, which is thick, and contains five rays.

From the beginning of the *back* fin to the mouth slopes suddenly, with an arch much more considerable than that from the mouth to the ventral fins. The *back* and *tail* have furrows for receiving the parts of the back and vent fins, that are supported by prickles. Near its end, the tail, both above and below, slopes very suddenly. The *lateral lines* are ill defined, but seem to be high, and to form an arch parallel to the back. The *scales* adhere firmly, and are large, smooth, and entire on the edge. Above each ventral fin is a short scale-like *appendant.* The *vent* is before the middle.

There is only one *back* fin, which rises gradually higher from the first prickle to the sixth soft ray, which is the longest, and then is rounded into a vertical end, the hinder rays gradually growing shorter to the last. The edge of the part supported by prickles is indented like a saw. The *pectoral* fins are low, short, and blunt; but their upper rays are the longest: each contains seventeen rays, of which the first is very short. The *ventral* fins are small and near each other, and each contains six rays, of which the first is a short prickle, and the others are branched. The fin behind the *vent* is shaped exactly like that of the back, the sixth soft ray being the longest. It contains thirteen strong prickles and eleven branched rays. The fin of the *tail* is shaped like a fan; but the middle of its hinder edge forms a very slight concavity. It contains sixteen distinct rays, besides some short compacted ones at the sides.

XI. Genus.—Pleuronectes.

Fishes of the fourth order, with both eyes on one side of the head, and with pectoral fins.

1st Species.—Pleuronectes nauphala.

A *pleuronectes* with eyes on the left side; with the fin of the tail sharp-pointed; with about seventy-three rays in the back fin, and fifty-five in that behind the vent; and with a long body devoid of tubercles on the upper side.

From the description of La Cepede, (*Hist. des Poissons*, Tome IV. p. 598,) it would appear, that the *nauphala* has some affinity to the *P. dentatus:* but the accounts of the latter are imperfect; and, from its being a native of North America, it may be presumed different from our Indian kind. This has also some affinity to the *P. rhombus*, or *pearl*, of the English; but is longer in the shape, like the *P. hypoglossus*, or *holybut*.

The *nauphala* is found in the Gangetic estuaries, grows to about eight inches in length, and was rejected with loathing by the natives, among whom I saw it. The fish is narrow in proportion to its length, being shaped like the iron of a lance, while its greatest width is before the middle. Its upper or left side is rough, and of a greenish-brown colour, on which are scattered some black spots. The under side is whitish and smooth.

The *head* is oval, sharp, small, rough, and covered with scales. The *mouth* is large, and, when shut, forms an oblique arch. The *jaws*, when open, are protruded, and the under one is the longest, and on each side without has a sharp ridge. The *lips* are somewhat fleshy, and the bones of the upper are dilated at the lower ends. In each jaw is a single row of sharp awl-shaped large *teeth*, nearly equal in size, and placed at equal distances from each other. The *tongue* is narrow, sharp, and small, and moves freely. The *palate* is smooth. The *nostrils* are in the middle between the eyes and jaw, each having two small apertures. The *eyes* are oval, with pupils of a kidney form. The *gill-covers* are scaly, somewhat arched behind, and move freely. Each consists of four plates, and conceals its *membrane*, which is thin, and contains seven rays inserted into a bone, serving as a support common to the whole.

The *lateral lines* run straight along the middle of the sides, but, in their fore part, send off a branch, which, forming an arch forward, is then bent up to the back. The *vent* is far before the middle. The *scales* adhere firmly; those on the upper side are finely indented on the edge, those below are smooth.

There are seven distinct *fins*. That of the *back* is slightly arched, and contains seventy-three undivided flexible rays, a little longer than the membrane. The *pectoral* fins are near the ventral, and are very small and sharp above. That on the left is rather the longest, and contains twelve rays, of which the third, fourth, and fifth, are bifid; while that on the right contains eleven undivided rays. The *ventral* fins surround that organ, and are very small. Each has six rays, of which the third and fourth are bifid, and the others undivided. The fin behind the *vent* is slightly arched, and contains fifty-five undivided flexible rays, a little longer than the membrane. The fin of the *tail* is undivided and sharp-pointed, and contains seventeen rays, of which two on each side are undivided, and the others are branched.

2d Species.—Pleuronectes arsius.

A *pleuronectes* with the eyes on the left; with the fin of the tail quite entire, with eighty-one rays in the back fin, and fifty-five in that behind the vent; with a long body; and with two eye-like spots on the upper lateral line.

This species has a strong affinity with the last, and also with the *Noree nalaka* of Dr Russell, (Indian Fishes, Vol. II. No. 77;) but this last may be readily distinguished by the number of rays in the fins. In Dr Russell's fish there are besides three eye-like spots; one on the lateral line, a second above, and a third below that organ.

The *arsius (arsi)* is a small fish which I found in the estuary below Calcutta. Its *shape* entirely resembles that of the *nauphala;* but it is smooth on both sides. The upper or left is of a greenish *colour*, with several brown cloud-like spots, and two that resemble eyes on the lateral line. The centre of these spots is black, and each is surrounded by an interrupted white circle. The under side is white. The fins are clouded.

The *head* is oval, sharp, small, and covered with scales. The *mouth* is large, and, when shut, forms an oblique arch. The *jaws* in opening are not protruded; the under one is the shortest. The bones of the upper *lip* are dilated at their lower ends. The *teeth* in both jaws are placed in a single row, taper to a sharp point, are large, and nearly equal in size, and stand at some distance from each other. On the left side there are two small *nostrils* in the middle between the jaw and eyes. The *eyes* are green and convex, with kidney-shaped pupils. The *gill-covers* are scaly, move freely, and conceal their *membranes*, each of which contains three or four rays.

The *lateral lines* behind are straight, and in the middle; over the pectoral fins they are arched, and, dividing on the head into two branches, which reunite at the nose, include a space which, on the left, contains the eye next the tail. On the right a third branch turns off to the back. The *vent* is very far forward. The *scales* are minute, and, so far as I can judge, smooth on the edge.

There are seven distinct *fins*. That of the *back* reaches from the eyes to the end of the tail, forming an arch. The *pectoral* fins placed near the middle of the sides are small and sharp, each containing ten rays. Each *ventral* fin has six rays. The edge of the fin behind the vent is arched. The *tail* fin is rounded behind, and contains sixteen rays.

3d Species.—PLEURONECTES PAN. Plate XXIV. Fig. 42.

A *pleuronectes* with the eyes on the right; with the fins of the back, tail, and vent, united into one; with a short body; and with five bristle-like rays in each pectoral fin.

This fish has all the marks mentioned by Linnæus in his specific character of the *Pleuronectes plagiusa;* yet it would be strange if the same species should be found inhabiting the mouths of rivers in India and Carolina, where the *P. plagiusa* lives. The accounts of this last fish in the *Tableau Encyclopedique*, and in La Cepede, (*Hist. des Poissons*, Tome IV. page 597,) are so short, that much reliance cannot be placed on what is there stated, although I see nothing very materially different in our fish from what is mentioned in the above works. The *P. plagiusa* is said to be oblong, but our fish is rather oval, that is to say, is wider than the term oblong would imply. Besides, the *Pàn* has pectoral fins of the singular structure observed in the *P. trichodactylus*, which derives its name from this circumstance; nor is it mentioned, that the fins of the *P. plagiusa* are of so remarkable a structure.

The *Pan* abounds in the eastern estuaries of the Ganges from Dhaka downwards, and is a delicious small fish, from four to six inches in length. It is nearly of an oval *form*, very much flattened, and rough on both sides, sweemming on the left, which is white. The upper side is of a brown *colour*, and marked with many dots, irregular black spots, and five or six transverse short lines falling at right angles on the upper side of the lateral line.

The *head* is blunt, small, very short, and covered with scales. At the right angle of the upper *lip* there is a short blunt *tendril.* The *mouth* is behind the nose, is small, and, when shut, forms a curve. The *jaws* do not protrude when open, and contain no *teeth.* The *lips* are fleshy; the upper one rather the longest and sharp; the under one has a slight notch at the end. The *palate* and *tongue* are smooth, and the latter is sharp. I can observe no *nostrils.* The *eyes* are convex and small, and even the irides are covered with scales. The *gill-covers* are simple, scaly, and rough, move freely, and their posterior edge forms a kind of arch, leaving a small opening at the throat, and concealing their *membranes,* each of which contains four rays.

The *lateral line* is straight, and in the middle. The *vent* is far before the middle. The *scales* are minute, oblong, and indented at the end with five or six little spines.

There are only five *fins,* those of the *back, tail,* and *vent,* being completely united into one, which extends from the vent to within less than a quarter of an inch from the mouth; nor are the parts distinguishable. The rays are shortest at both terminations, and lengthen gradually to the centre of the tail, which is rather sharp. The whole fin is scaly, and has one hundred and fifteen undivided rays, of which sixty-four are above, and fifty-one below. The *pectoral* fins are near the vent, blunt and minute, each containing five hair-like rays, of which the second is the longest. The *ventral* fins are very small, surround the vent, and approach very near, each containing four rays.

XII. Genus.—Achirus.

Fishes of the fourth order, with both eyes placed on one side, and destitute of pectoral fins.

1st Species.—Achirus cynoglossus.

An *achirus* with the eyes on the left; with the fins of the back, tail, and vent united, but distinguishable; the first containing one hundred and two rays, the second twelve, and the third seventy-eight; and with the left side dotted with brown.

The only species of this genus, which I have found in the Ganges, is usually called a *Sole* by the English, on account of its having very exactly the excellent taste of that valuable fish. The native name being somewhat harsh, I have translated it into the dialect of Greece. It has the utmost affinity to the *Achirus bilineatus* of La Cepede, (*Hist. des Poissons*, Tome IV. p. 659,) which, in its quality, is equally valuable. It has also a strong resemblance to the *Jeree potoo* of Dr Russell, (Indian Fishes, Vol. I. No. 73,) which differs in the number of rays supporting the fins. Nor is the resemblance less strong between it and the *Cynoglossus lingua*, described in the Second Order, (p. 32.) Indeed, the two species are so like, as to demonstrate, that the arrangement, founded on the absence or situation of the fins, however convenient, is totally unnatural.

This fish is found in the mouths of the Ganges as high as the tide reaches, but is not common, except where the water is strongly impregnated with salt, and seldom grows to above six inches in length.

The *Cynoglossus* sweems on the right side, is very flat, of a long oval *form*, and covered with scales even on the head. The left, or upper side, is of an uniform brown *colour*, dotted towards the edges. The right is white. The fins are brown, dotted with black.

The *head*, viewed from above, is oval, small, and blunt. The *mouth* is small, placed at some distance below the extremity of the nose, and passes very obliquely towards the eyes. The *lips* are fleshy; the upper one is the longest, and is incurved, sharp, and without bones; the under one is blunt. There are no *teeth*, except a very slight roughness on the jaws. I have not been able to observe any *nostril*, unless it be concealed in a small membrane proceeding from the upper jaw, and below the under eye. The *eyes* are small. The *gill-covers* are flexible, and move freely. Their *membranes* contain no rays.

The *lateral line*, on the left side, is double; the superior running near the back, and the inferior straight along the middle. On the right the lateral line is bifid, with the branches separating wide, one running towards the back, and the other near the belly. The *vent* is immediately behind the throat, inclining towards the right, or under side. The *scales* are indented on the edge, and adhere firmly. The *tail* is blunt.

There are properly only two *fins*: one runs from the crown of the head, round the tail, to the vent; but the rays of the *tail* fin are longer than those of the *back* and *vent* fins; and thus the three may be distinguished, although the fins of the back and vent are joined to that of the tail for the whole length of their last rays. There is only one *ventral* fin, which is very small, and placed longitudinally towards the left side, opposite to the vent. It contains four soft rays.

XIII. Genus.—Calliomorus.

Fishes of the fourth order, with the head armed with spines, but with the edges of its bones unindented; with the body depressed, and covered with scales; and with two back fins.

1st Species.—Calliomorus chaca.

A *calliomorus* with a spine before each eye, and with the hinder bone of the gill-covers unarmed.

If this fish be a distinct species from the *Calliomorus indicus* of La Cepede, (*Hist. des Poissons*, Tome II. p. 344,) as the differences in the position of the spines on the head, and in the number of rays, would seem to indicate, it has the utmost affinity to that fish, which is the *Callionymus indicus* of Linnæus, Bonnaterre, (*Tabl. Encycl.* p. 44,) and Russell, (Indian Fishes, Vol. I. No. 46.) Bloch was sensible of its belonging to a very different genus from the *Callionymus*, and called it *Platycephalus spathula*, (*Ichth.* Tome XII. p. 90, Planche CCCCXXIV.) La Cepede was of the same opinion; but, with his usual desire of appearing new, changed Bloch's name into one more objectionable, and gave it a wrong place in his system; for, as Dr Russell justly observes, the ventral fins are not placed before, but opposite to the pectorals; and, although it has an affinity with the *Callionymus Uranoscopus*, and *Trachinus*, as Linnæus rightly observes, it also strongly resembles the *Cottus*. It has also some resemblance to the genus *Gobius;* and one of its native names signifies that it is a flattened *Gobius;* while another that I have adopted denotes its affinity to the next fish that will be described, although that belongs to the order of abdominal fishes.

The *Chaca* is a rough long flattened fish, with its tail ending like a wedge. Above it is of a green *colour*, clouded with broad irregular black bars. Beneath it is whitish. The fins of the back and tail are variegated with black.

The *head* is blunt, wedge-shaped, a little wider than the body, horizontal above, much depressed, rough, and covered with scales. Along it run nine *ridges*, of which that in the midst, after a short interruption, is divided into three at its fore part; the two next run from above the eyes to the nape; the two next from the posterior angles of the eyes to the beginning of the lateral lines; and two run from each corner of the mouth to the end of the cheek-bones. Above the anterior angle of each eye is a recumbent strong short *prickle*. At the posterior angle of each cheek-bone are two strong awl-shaped *thorns*. The *mouth* is large, and descends obliquely in a straight line. The *jaws*, when open, are protruded; and the upper, which is the shortest, has a ridge in the middle. The *lips* are fleshy, the upper one double, and without bones. On the lower jaw is a line covered with man crowded *teeth*. In the upper jaw is a similar line, which, towards each corner, grows narrow; and in the middle, among the small sharp crowded teeth, are intermixed some large ones. On the fore part of the *palate* is a bone in form of a crescent, and covered with teeth; and behind this, on each side, is a row of erect parallel teeth, placed close to each other. The *tongue* is smooth, broad, and ends in a concavity. The *eyes* are on the crown of the head, but far from each other, and far forward. They are small, with heart-shaped pupils; and the animal possesses the power of withdrawing them to the level of the head, and of then again rendering them protuberant. The *gill-covers* consist of one plate, rounded behind, with a sinus above and below. They move freely, leave a large opening, are covered with scales and rough, and do not conceal their *membranes*, each of which contains seven rays.

The *back* is straight, rounded across, and provided with a furrow for lodging the foremost back fin. The *lateral lines* are high, straight, and elevated to a ridge. The *vent* is before the middle. The *scales* are small, indented on the edge, and of an oblong form, and adhere firmly.

On the *back* there are two distinct *fins*. The first contains eight prickles, and slopes down from the third, which is the longest, the two first being very short. The second fin is indented on the edge, and contains thirteen rays, of nearly equal length; the first being undivided, and the others branched.

The *pectoral* fins, placed near the middle of the side, are short, and rounded at the end: each contains eighteen rays, of which two on each side are undivided. The *ventral* fins are inserted far from each other, on two naked bones, which form the lower edge of the body. Each contains six rays, of which the hindmost are bifid and the longest, while those before are branched. The fin behind the *vent* is indented on the edge, and contains thirteen branched rays, nearly of equal length. That of the *tail* is rounded behind, and contains about eighteen rays.

ORDER V.

FISHES WITH THE DORSAL SPINE OF BONE, AND HAVING THEIR VENTRAL FINS BEHIND THE PECTORALS.

SEVERAL *genera* belonging to this order, although often distinct enough in general appearance, are very difficult to be distinguished by well-defined characters. The most careless observer, for instance, would have little difficulty in distinguishing the species belonging to the genus *Cobitis* from those constituting the genus *Cyprinus*, or the genus *Silurus;* yet I must say, that I find it almost impossible to state any well-defined marks, by which the *Cobitis* can be distinguished, on one hand, from the *Cyprinus*, and, on the other, from the *Silurus*. But these genera must be considered by all as the fishes of this order most distinct from each other; the Linnæan genera *Salmo*, *Mugil*, *Atherina*, *Esox*, and *Clupea*, being interposed in the scale of gradation. It must be farther observed, that different species of *Cyprinus* have very strong affinities with each of these intermediate *genera*, while the *Silurus* has very little affinity with any of them, except the *Salmo*. In this view, the whole of these *genera*, with their subdivisions, as detailed by La Cepede, forms a connected circle, which, contrary to the order of writing, has neither beginning nor end. In treating of them, therefore, systematically, the naturalist, with equal propriety, may begin with any genus; but he must finish with one nearly allied to the genus with which he commenced. This will serve my purpose, as including all the *genera* of this order which I have observed in the Ganges, although authors describe several that are quite anomalous, and that have no affinity to any of those connected in the circle.

I begin with the *genera*, comprehended by Linnæus under the term *Silurus*, very properly subdivided by La Cepede, as comprehending too many species. The fishes of this kind may be distinguished by having a very slimy skin, destitute of scales, and by having either two fins on the back; or, if they have only one fin, it is either united to that of the tail, or is destitute of rays, or placed far forward, or is very long. The genus *Cobitis*, allied to these fishes, has only one short fin on the back, and it is supported by rays, and placed either on the middle of the back, or immediately behind it.

Besides these general characteristic marks, there are some peculiarities to be observed in many of the species, although all are not common to every species.

None of these fishes have the bones in the upper lip called *mystaces* by some authors. Some, indeed, in place of these bones, have small cartilaginous elongations, but they are of such trifling dimensions as to attract little or no notice. In none of these fishes do the jaws protrude in opening, and all of them have two apertures in each nostril. All these fishes have moveable gill-covers, leaving a large opening on each side.

In many species, there are one or more bones on the palate adjacent to the jaw, and, like this, covered with numerous crowded teeth. And the same is the case with four bones, two above and two below, which many species have at the entrance into the throat. In a few species, there are either no teeth, even in the jaws, or these organs are extremely minute.

Although the surface of these fishes has always an uncommon number of pores for the secretion of slime, and although they are destitute of scales properly so called, some of them have their sides more or less cased in bony plates; but, as none of these are among the fishes here described, I shall not dwell any farther on this subject. Many species, however, have no skin on considerable portions of the bony plate, which covers the head, and sends processes along the shoulders and fore part of the back; nor on a bone, which is placed before the first fin of the back; nor on a bone analogous to our *scapulæ* or shoulder-blades, which extends along the side above the pectoral fin; nor, finally, on the bone of the breast or sternum. All of these bones, instead of being concealed under the skin, have no covering but a thin cuticle, through which they are distinctly visible, and they are usually marked by various elevated ridges, and numerous waved lines intersecting each other in different directions, or disposed as if radiating from different centres.

Under the lower jaw, and before the *os hyoides* on the throat, many of these fishes have a transverse slit, resembling somewhat a second mouth. It usually forms an angle or curve, with the two extremities turned backward. The slit penetrates from a tenth to a half inch in depth, in proportion to the size of the fish; but is lined by a smooth cuticle, nor has it any communication with the throat, mouth, or any considerable gland; nor, so far as I could observe, does it ever enlarge within into any thing like a hollow sack. This slit, although somewhat differently situated, has an affinity with the subocular sack of some species of the ruminant quadrupeds, although I cannot state the use of such an organ in either class of animals.

Several species have two apertures in the vent, one probably an organ of procreation, and the other of digestion. I do not, however, know that both sexes of the same species have this double aperture; and suspect that it belongs only to the male, as I have several times observed a fleshy substance protruded from the foremost aperture, which I considered to be a kind of *penis*, although I never ascertained the point by dissection; nor did I notice it in every individual that I observed with two apertures.

A great many species are provided with a strong prickle in the first fin of the back, and in each pectoral fin. These prickles are no doubt formidable weapons of defence, as they are capable of inflicting severe wounds on any enemy that attempted to seize the animal which they guard; yet in general, every time that they are employed, the animal must suffer considerably, as, in most of the species, these prickles terminate in a flexible substance resembling somewhat whalebone; and, before the prickle can penetrate an enemy, this flexible point, which supports part of the membrane, must be broken. Although, therefore, it is probable, that, in case of such accidents, the flexible part may be soon reunited to the prickle; yet it is not likely that the animal should have recourse to the use of such a weapon, except in emergency, and perhaps never as a weapon of attack. These weapons are indeed so formidable, not only from their strength, but from their being in general barbed on the edge, so as to occasion very painful punctures, that every enemy probably attacks with great caution, and that few even ever make the attempt, except such as, from an enormous difference in strength, art, or organization, are little exposed to danger. Man, therefore, is almost the only enemy which most species of these fishes have to dread.

I. Genus.—Platystacus.

Fishes of the fifth order, with a very slimy skin, destitute of scales, but provided with numerous soft projections; with a mouth surrounded by tendrils; with one short back fin placed before the middle; with the tail much shorter than the trunk, and not compressed; and with prickles in the pectoral and dorsal fins.

The genus *Platystacus* constituted by Bloch consisted of two species, but has not been adopted by La Cepede. One species he has united with the *Silurus*, while the remaining species, which he could not unite with that genus, he calls *Plotosus.* As I think that the species which he has placed in the genus *Silurus* not only differs very much in appearance from those placed in the last mentioned genus, I shall retain Bloch's name, Platystacus, and apply it to the *Silurus verrucosus* of La Cepede, (*Hist. des Poissons*, Tome V. p. 56,) or the *Platystacus verrucosus* of Bloch, (*Ichth.* Tome XI. p. 61, Pl. LXXIII. fig. 3,) and to the species which I am now about to describe. This genus has much of the manners of the *Cottus uranoscopus*, and other lurking fishes.

1st Species.—Platystacus chaca. Plate XXVIII. Fig. 43.

A *platystacus* with the fin of the tail extending much farther on the upper than on the under edge of the tail.

The *Chaca* is found in the rivers and ponds of the north-eastern parts of Bengal; grows to about a span in length; but is so ugly, that all reject it for food with loathing. It is of a depressed *form*, ending behind in a tapering tail. All over the skin are scattered little fleshy projections, ragged on the surface. Above it is clouded with green and black, and below with the latter *colour*, and yellow; but all its colours are dirty and ill defined. The fins are spotted with black.

The *head* is much flattened, very blunt, and wider than the body. Down its middle runs an oblong cavity, surrounded by an ill defined ridge, from which there proceed nine others, diverging as if from a common centre. There are six *tendrils*, much shorter than the head, four from the sides of the under jaw, and from each corner of the mouth one, which is connected with a short membrane indented with three points. The *mouth* is terminal, but high, and of enormous size. The under *jaw* is the longest, and is provided with a membranous veil. The *lips* are very thick, and indented on the edge. In both jaws are crowded many sharp minute *teeth*, those in the upper jaw being confined to two long spaces. On the palate are four spaces covered with similar teeth. The *tongue* is very broad, and has a notch at the end. Each *nostril* consists of a minute tube on the edge of the lip. The *eyes* are forward on the crown of the head, but far from each other, and are small, protuberant, and covered by the common skin. The pupils resemble slits. Each *gill-cover* consists of a single plate, which does not conceal the membrane; but this is so thick, that no rays can be distinguished. The skin under the lower jaw does not form a sack.

The *body* is depressed, and the *tail* tapers without being compressed. The *lateral lines* descend along the sides, and are smooth. The *vent* is behind the middle.

The *fins* are covered with a thick skin. That of the *back* is very short, and rather before the middle. It contains a strong sharp prickle, indented on the fore side, and three soft rays. The *pectoral* fins are small and rounded, each having about nine rays, of which the first is a very strong sharp prickle, having the outer edge covered with tubercles. The *ventral* fins are larger than the pectorals, and remote from each other; a little behind the back fin. Each contains eight rays. The fin, placed half way between the *vent* and end of the tail, is very short and rounded, and contains eight rays. The *tail* fin, on the upper side, extends beyond the vent fin, but below reaches only near it. It is rounded at the extremity, and contains about thirty-eight rays.

II. Genus.—Plotosus.

Fishes of the fifth order, with a smooth slimy skin, destitute of scales, with a mouth surrounded by tendrils, with the hinder back, tail, and vent fins, united; and with the tail much longer than the head and body.

1st Species.—Plotosus canius.

A *plotosus* with the back of an uniform colour, and with the prickle of the back fin smooth.

Plotosus anguillaris, La Cepede, *Hist. des Poissons*, Tome V. p. 130?

Platystacus anguillaris, Bloch, *Ichth.* Tome XI. p. 49, Pl. CCCLXXIII. fig. 1?

Ingeelee, Russell's Indian Fishes, Vol. II. No. 166?

In the fishes described by these three authors, the sides are marked by longitudinal white stripes, which are wanting in the *Canius*. The two first authors mention five stripes in all on a purplish brown ground; while they state, that the prickle in the back fin is smooth, and the tendrils longer than the head, as in our fish; but theirs has a much longer and slenderer tail than the *Canius.* Russell again mentions, that the *Ingellee* on each side has three stripes on a copper-coloured ground, the prickle of the back fin indented, and the tendrils shorter by one half than the head. I am therefore inclined to believe, that there are three species of this genus, which is exceedingly well marked, and totally different in its habits from the *Platystacus,* although very nearly allied to the three fishes that will be next described.

The *Canius (Kani magur)* is found in the rivers of the southern parts of Bengal; and, although a very ugly animal, and rejected by Europeans, is considered by the natives as excellent food. It is often found three feet long, and sometimes between four and five. It is a long fish, with somewhat of the eel *form,* although not quite so flexible, and is of a lurid uniform dark olive *colour,* with a tinge of violet. It is quite smooth, slippery, and devoid of scales.

The *head* is long, flattened, rather wider than the body, very blunt, and smooth. Round the mouth are eight awl-shaped *tendrils,* of which the six uppermost are nearly the length of the head, and the two under the chin are a little shorter. The *mouth* is of a moderate size, and the upper *jaw* is the longest. The *lips* are fleshy. The foremost aperture of each *nostril* is adjacent to the upper lip, and tubular; and the hinder, in the middle between the eyes and nose, is oblong. The *eyes,* placed on the crown of the head at a distance from each other, are rather small, and of an oblong form.

The fore part of each *lateral line* ascends a little on the shoulder. The *vent* is near the middle. The *trunk* is short and rounded; the *tail* is compressed, ending in a sharp point.

The six *fins* are covered with so thick a skin, that the number of rays which each contains could not be ascertained. The fore fin of the *back,* placed on the middle of the trunk, is very short, and somewhat triangular. It contains a very strong sharp prickle, smooth on the edges. The *pectoral* fins are low, small, and sharp above. Each contains a very sharp strong prickle, indented backward on the outer edge. The *ventral* fins are very small, and rounded. That of the *tail* extends above, and below the whole length of that organ, and is evidently composed of a posterior back fin, united with those of the tail, and behind the vent, but no separation of parts can be observed: its extremity is sharpish.

III. Genus.—Macropteronotus.

Fishes of the fifth order, with a smooth slimy skin, destitute of scales; with the mouth surrounded by tendrils; with one long fin on the back; with prickles in the pectoral fins; and with the tail wedge-shaped.

The two species of this genus, that I have seen, are found in ponds and ditches, and, as these dry up, are caught nestling among the mud. They are very ugly dirty-looking animals, on which account, probably, they are seldom, if ever, used by Europeans; but the natives consider them as wholesome invigorating diet, and employ a broth made of them in curing those troubled with pains in the abdomen. They grow from a foot to a foot and a half in length. Besides the circumstances mentioned in the generic character, these fishes have in common the following appearances:

Although shorter in proportion, less flexible, and more flattened at the tail, they approach a good deal to the eel in *form.*

The *head* is blunt, depressed, broader than the body, but short and smooth. The mouth is surrounded by eight awl-shaped *tendrils*, of which some are a little longer, and others a little shorter than the head. The *mouth* extends directly backward, and is rather small. The under *jaw* is shortest. The *lips* are fleshy, and the upper has no bones. In both jaws are numerous crowded sharp granular *teeth.* The *tongue* is blunt, smooth, and tied down. The foremost aperture of each *nostril,* placed at the end of the nose, is tubular; the hinder one is half way between the former and the eye. The *eyes* are placed forward on the crown of the head, but far from each other, are small and round, with circular pupils. The *gill-covers* move freely, consist of one plate covered with skin, smooth, and ending in a point behind. The *membranes* of the gill-covers are exposed to view, but are so thick, that the number of rays, which each contains, cannot be exactly determined. Under the lower jaw is a transverse slit.

The *sternum,* or breast-bone, is naked, and marked with transverse lines like the teeth of a comb. The *lateral lines* run straight along the middle of the side, and are smooth. The *vent* is before the middle.

There are seven *fins* covered with thick skin, but their rays are visible. That of the *back* reaches from very near the head to very near the end of the tail. Its rays, which are soft, branched, and recurved at the end, gradually lengthen a little toward the posterior end, which is rounded. The *pectoral* fins are low, small, and blunt, although their rays are longest above the middle. The first ray in each is a strong sharp prickle, not indented on the edge. The *ventral* fins are small and near each other, and each contains six soft rays. The fin behind the *vent* extends from near that organ to the end of the tail, and resembles that of the back. The tail fin is rounded.

1st Species.—Macropteronotus jagur.

A *macropteronotus* with the back fin united to that of the tail.

This species is rather the largest. Above it is of a blackish green *colour*, and below whitish, while the sides are slightly variegated with clouds. The fins are of the same colour with the parts on which they are inserted.

On the crown of the *head* is one oval concavity. The two *tendrils* at the hinder apertures of the nostrils are as long as the head; the two at the angles of the mouth are a little longer; of the four under the chin, the two outer ones are a little longer than the inner, and reach to the back of the head. The *palate* has no teeth. The *eyes* are convex. I can observe two strong rays in each *membrane* of the gill-covers.

Although the *back* fin is connected with that of the *tail*, the two are easily distinguishable, the former containing fifty-three rays, and the latter about fourteen. Each *pectoral* fin has seven branched rays and the prickle. Each *ventral* fin has four branched rays, with an undivided one at each side. The fin behind the *vent* contains fifty rays.

2d Species.—Macropteronotus magur. Plate XXVI. Fig. 45.

A *macropteronctus* with the tail and back fins distinct; with sixty rays in the former, and forty-four in that behind the vent; and with the edge of the pectoral prickles unindented.

This species, if it be distinct, has the utmost affinity to the *Macropteronotus batrachus* of La Cepede, (*Hist. des Poissons*, Tome V. page 84,) or the *Silurus batrachus* of Bloch, (*Ichth.* Tome XI. page 34, Pl. CCCLXX. fig. 1.) The fish, indeed, described by these authors differs from ours in the number of rays supporting its fins, but, owing to the thickness of the skin, no great reliance can be placed on the accuracy of this mark; and besides, these authors describe their fish as of an uniform brown colour, and state, that the prickle of each pectoral fin is strongly indented behind.

The *Magur* is *shaped* somewhat like the head of a lance, and is of an olive green *colour* above, and dirty yellow below. Its fins are edged with red.

The crown of the *head* is covered by a naked bone, in which are impressed many hollow dots and two concavities; the foremost, between the eyes, is long; and the other, above the pectoral fins, is circular. The two *tendrils* at the nostrils are shorter than the head; the two at the corners of the mouth are almost twice as long as the head; and of the four under the chin, the two middle ones are as long as the head, and the two outer ones are a little longer. In the fore part of the palate is a crescent-shaped bone covered with *teeth.* In each *membrane* of the gill-covers I reckoned nine rays, without being sure of having seen the whole.

There are one prickle and nine soft rays in each *pectoral* fin, the former being very short and strong. Each *ventral* fin has one undivided and five branching rays. The fin of the tail fifteen, of which one on each side is undivided.

IV. Genus.—Silurus.

Fishes of the fifth order, with a smooth slimy skin, destitute of scales; with the mouth surrounded by tendrils; with a long and compressed tail; with one very short back fin, placed on the shoulders, and supported by rays; and either with prickles in the back or pectoral fins, or with more than three rays in the membranes of the gill-covers.

1st Species.—Silurus singio. Plate XXXVII. Fig. 46.

A *silurus* with a rounded tail fin; with eight tendrils; and with the prickles of the pectoral fins barbed on both sides.

This fish, if really different, is most nearly allied to the *Silurus fossilis* of Bloch, (*Ichth.* Tome XI. p. 36, Pl. CCCLX.

fig. 2,) and La Cepede, (*Hist. des Poissons*, Tome V. p. 74.) Their fish is of a chocolate colour; its lateral lines ascend towards the shoulder; and the number of rays in its fins is somewhat different from that of the fish in Bengal. The most marked differences, however, to judge from the figure in Bloch, are, that he represents the prickles of the pectoral fins as slender, and without barbs.

The *Singio* (*Singgi*) resembles, in manners, appearance, and qualities, the three last described fishes; but, being reckoned highly nutritious, so that it is eagerly sought after by nurses, and by men worn out with too great an indulgence of their passions, it is seldom found fully grown. On account of its qualities, perhaps, many of the Pagans abstain from it, when about to perform any religious ceremony.

The *Singio* is occasionally found a foot and a half in length; but a foot is its more ordinary dimensions. It has much of the eel in its *form*, and is of a dirty green *colour*, with reddish fins. On each side it has a faint yellowish longitudinal stripe above the lateral line, and another below that organ.

The *head* is very blunt, slopes a little toward the mouth, and is small, depressed, and covered by a smooth bone, which terminates behind in three sharp points, of which the middle one is by much the smallest. On its summit are two oblong cavities, the foremost of which, placed between the eyes, is the longest. All the *tendrils* are awl-shaped, and longer than the head: the two from the corners of the mouth, and the four from the chin, extend to about the end of the pectoral fins; but the two at the hinder nostrils are shorter than the others. The *mouth* is of a moderate size, and extends directly backward. The *jaws* are nearly equal in length. The *lips* are fleshy. In each jaw are crowded many minute sharp *teeth*. The *tongue* is blunt, and smooth. The foremost aperture of each *nostril*, situated at the lip, is tubular; the hinder one is oblong, and in the middle between the eye and jaw. The *eyes* are small, and high. Each of the *gill-covers* consists of one plate, very sharp behind. Its *membrane* is fully exposed on the throat, but is so thick, that its rays, if it contains any, are not visible. Under the lower jaw is a *slit.*

The *sternum* is exposed to view, and is striated with converging lines. The *lateral line* passes straight along the middle of the side, but is faint. The *vent* is far forward. The *tail* forms a sharp ridge above and below.

The *back* fin is rounded at the end, and contains seven rays, of which the first is simple, but not a prickle: the others are branched. The *pectoral* fins are low and blunt, but longest above: each contains eight rays, of which the first is a strong sharp prickle, indented on both edges with barbs pointing downward. The *ventral* fins are small, approach near each other, and each contains six branched rays. The fin behind the *vent* reaches close to that of the tail, but is not united to it, and, forming a slight arch with its edge, contains about sixty-one rays; but the skin is so thick, that the number of rays cannot be satisfactorily ascertained. The fin of the *tail* is rounded, turned obliquely downward, and contains about twelve rays, besides some short ones at each side.

Subgenus.—SILURUS CALLICHROUS.

The following five species for this tribe of fishes are rather handsome, and have little or nothing of that lurid appearance by which many kindred species are distinguished. They are all very rich fine-flavoured food, and grow to from nine to twelve inches in length. As they all have the strongest resemblance to each other, I shall content myself with describing the first species only at length, and shall afterwards point out the chief circumstances in which the others differ. In the first four species I did not ascertain whether or not the jaws protrude in opening; in the fifth species they do not.

2d Species.—SILURUS PABDA. Plate XXV. Fig. 47.

A *silurus* with the fin of the tail divided into two blunt lobes, with four tendrils, of which the two upper reach beyond the pectoral fins; with the prickle in each of these smooth on the edges; with fifty-four rays in the fin behind the vent; and with the sides clouded with irregular black marks.

This fish is found in the ponds and rivers of Bengal. It is of a long lance-shaped *form*, and much compressed at the sides. The *colour* above is greenish, and below white, with a longitudinal stripe of yellow, changing to purple, above the lateral line. The clouds on the sides consist of black dots. Above each pectoral fin is an irregular dark mark peculiar to this species. The sides are somewhat diaphanous, so that the divisions of the muscles can be perceived.

The *head* is blunt, oval, slightly flattened, and a little wider than the body, but very short. The upper *tendrils* proceed from above the nostrils; the lower from under the chin; and these are shorter than the head. The *mouth* is large, and slopes a little downward, the upper *jaw* being considerably the shortest. The *lips* are fleshy. In both jaws are crowded many small unequal reflected sharp *teeth;* and there are some similar on the palate. The *tongue* is blunt, thick, smooth, and marked transversely by three pair of veins. The apertures of the *nostrils* are round, and in the middle, between the eyes and jaw, but above the former. The *eyes*, which are small and round, are on the side of the head, and near its middle. Each *gill-cover* consists of one sharp plate, and leaves exposed to view a transparent membrane, containing nine rays. Under the lower jaw is a *slit.*

The *back* slopes from the fin to the head, and the *belly* projects as much. The *lateral line* runs straight from the upper angle of the gill-covers to the end of the tail, and is smooth. The *vent* is plaeed very far forward. The *tail* is straight, tapers gradually, and is rounded above, with a sharp edge below. No *naked* bony plates are exposed to view.

The *back* fin contains four soft rays, of which the first is not divided. The *pectoral* fins are low and short, but are longest above the middle, although they are rather blunt: each contains twelve rays, of which the first is a smooth, rather slender prickle, with its extremity of a substance resembling whalebone. The *ventral* fins are minute, and near each other, and each has eight rays. The fin behind the *vent* reaches close to that of the tail, but is not joined to it; and its rays are all soft, and of nearly equal lengths. The *tail* fin contains seventeen rays.

3d Species.—SILURUS CANIO.

A *silurus* with the fin of the tail divided into two sharp lobes; with four tendrils, of which the two upper reach almost to the middle of the fish; with sixty-nine rays in the fin of the vent; with the prickle of each pectoral fin smooth-edged; and with the sides unspotted.

This fish has a very strong resemblance to the *Silurus bimaculatus* of Bloch, (*Ichth.* Tome XI. p. 17, Pl. CCCLXIV.) and of La Cepede, (*Hist. des Poissons*, Tome V. p. 57;) but the tips of its tail fin are not black, a circumstance to which Bloch's fish owes its name. Besides, in Bloch's fish the first ray of each pectoral fin is a very strong indented prickle.

The *Canio* (*Kani pabda*) I found in ponds in the north-east parts of Bengal. It differs from the *Pabda* in having no clouds on its sides.

Behind the upper jaw there are two bones covered with *teeth;* and in each *membrane* of the gill-covers there are about twelve rays.

Each *pectoral fin* contains thirteen rays; the first is only hard at the very bottom. The fin of the *tail* contains eighteen rays.

4th Species.—SILURUS DUDA.

A *silurus* with the fin of the tail divided into two sharp lobes; with four *tendrils,* of which the two upper reach almost to the middle of the fish; with seventy-three rays in the fin behind the vent; and with the first ray of each pectoral fin smooth-edged.

The *Duda* (*Dudha pabda*) differs in nothing from the *Canio* except in the number of rays in the fin behind the vent; but the fishermen allege that they are different species. It is found in the river Kusi.

5th Species.—SILURUS CHECHRA.

A *silurus* with the fin of the tail divided into two blunt lobes; with four tendrils, of which the two upper reach to the end of the back fin; with sixty-seven rays in the fin behind the vent; with the prickle in each pectoral fin indented behind; and with the sides marked with cloud-like spots.

This species I also found in the Kusi river. The general *colour* is silver, with some dusky green on the back, and cloud-like spots on the sides, but without the yellow stripe along the sides that the *pabda* has. On each pectoral fin is a large black spot.

Like the *Canio,* this has a bone covered with *teeth* on each side of the palate, contiguous to the jaw. The rays in the *membrane* of each gill-cover are about thirteen.

The first ray of the *back* fin is almost twice as long as any of the others. The first ray of each *pectoral* fin is a strong prickle, indented on the hinder edge. The rays in the *vent* fin are about sixty-seven in number, but I am not satisfied that this was exactly determined.

6th Species.—Silurus pabo. Plate XVII. Fig. 48.

A *silurus* with the fin of the tail divided into two lobes, with four tendrils all shorter than the head, and with seventy-three rays in the fin of the vent.

This fish I found in the Brahmaputra river, towards Asam. It differs more from the *Pabda* than the intermediate species; but still a full description will be unnecessary. Above it is of a purple *colour*, with a green gloss; below the colour is silver, with a purple gloss. The belly is of a livid red. The head and sides are dotted with black.

The upper *tendrils* are close to the middle of the upper jaw, and reach only to the eye. There are two small bones in the upper *lip*. On the palate are six bones covered with *teeth*, two adjacent to the jaw, and four at the throat. The *tongue* wants the transverse marks, which that of the *Pabda* has. The *gill-covers* are rounded behind, and their membranes contain each about thirteen rays. There is no slit under the throat.

The *back* slopes with a considerable convexity. The *lateral line* forms an arch above the pectoral fin. The end of the tail is bent downward.

In the *back* fin are five rays, and sixteen in each *pectoral*, the hind part of the prickle being indented on the edge, while it is much stronger than in the *Pabda*; but, as in that fish, it terminates in a substance like whalebone. Each *ventral* fin contains ten rays. The fin behind the *vent* is separated by half an inch at least from that of the tail, and is covered half its height with scales, a very rare occurrence in this genus. The number of its rays is about seventy-three; but I cannot count them with certainty. The two lobes of the *tail* fin have their points turned much from each other, and are rather sharp. The whole contains nineteen rays.

7th Species.—SILURUS BOALIS. Plate XXIX. Fig. 49.

A *Silurus* with the fin of the tail divided into two lobes, with four tendrils, of which the two upper extend to the middle of the fish, with the pectoral fins unarmed, and with the edges of the first ray in the back fin smooth.

The *Wallagoo* of Dr Russell (*Indian Fishes*, Vol. II. No. 165) differs little from the *Boalis;* but has ninety-one rays in the fin of the vent, a mouth which descends obliquely, nostrils near the nose, lateral lines running parallel to the back, unspotted sides, and the first ray of the back fin strongly indented on the edge.

The *Boalis* is a very common fish in the rivers not only of the Gangetic provinces, but all over India, and is occasionally found in ponds. It grows to a large size, being sometimes found six feet in length, and is considered good eating by the natives; but has such a forbidding aspect, that few, if any, Europeans have ventured to taste it. The appearance of the fish is much more disgusting than the figure represents; yet, in other respects, it has a strong resemblance to the five preceding species. It is not, however, so high in *proportion* to its length, the slope from the back fin to the end of the nose being small and uniform all the way. Above it is of an olive *shade* with a golden gloss, and below white with clouded spots, consisting of numerous black dots; but the vent and tail fins are blackish, and

1

over the whole fish is spread a livid hue, which rapidly increases after it is taken out of the water, and soon covers the whole.

The *head* is short, blunt, flattened, and wider than the body. The upper *tendrils* arise from the middle of the upper jaw: the under ones from the throat, and these are shorter than the head. The *mouth* is immense, and extends straight back from the end of the nose to behind the eyes. The under *jaw* is rather the longest. The *lips* are scanty. The *teeth* in both jaws, and in a line parallel to each side of the upper on the palate, are crowded in great numbers, sharp, and reflected. The *tongue* is tied down, smooth, cartilaginous, and blunt. The apertures of the *nostrils* are round, and placed in a hollow line in the middle between the eyes and jaw. The *eyes* are high and small, with circular pupils. Each *gill-cover* consists of one rounded plate, which leaves exposed a thick *membrane*, containing nineteen rays. On the throat is a *slit*.

The *lateral line* is arched before, runs high along the side, and behind is smooth and porous. The *vent* is before the middle. The *tail* ends in a notch.

The fin of the *back* is indented on the edge, and contains five soft rays, diminishing rapidly in length from the first, which is undivided, and smooth on the edges, but jointed. The *pectoral* fins, which are short, low, and sharp above, have each fifteen soft jointed rays, of which the first, and perhaps the last, are undivided. The *ventral* fins are very small, and nearly approximated; and each has ten soft rays, of which the first is undivided. The fin behind the *vent* is rounded behind, and quite distinct from that of the tail, although it reaches very near it. The number of its rays, all nearly equal in length, is eighty-six, of which the first and second are undivided, and short. The *tail* fin is divided into two unequal lobes, the upper being longest, and contains seventeen distinct rays, with two minute ones at each side.

8th Species.—SILURUS GARUA. Plate XXI. Fig. 50.

A *Silurus* with the fin of the tail divided into two lobes; with eight tendrils; and with eight rays in the back fin, and twenty-nine in that of the vent.

The *Garua* is common in the fresh water rivers of the Gangetic provinces, often grows to two feet in length, and by the natives is considered as good eating. It has little or none of that lurid appearance to which many kindred fishes are liable, but differs more than the *Boalis* from the *Pabda*. The prevailing *colour* is silver, with green on the back. The fins are diaphanous, that on the back and the pectorals being dotted, and that of the tail being edged with black. The head and shoulders have a golden gloss. The fish is compressed at the sides, and of a long lance *shape*, tapering very gradually both to head and tail.

The *head* is oval, blunt, of a moderate size, and rather wider than the body, but by no means flattened above. Between the eyes there is a longitudinal cavity extending to the nostrils. There are eight *tendrils*: one pair, from above the hinder apertures of the nostrils, extend only to the eyes; the second pair, from the sides of the upper jaw, extend to the end of the ventral fins; the other two pair, coming from the chin, extend to the end of the head. The *mouth*, placed a little below the end of the nose, is large, and extends straight backward. The upper *jaw* is rather the longest. The upper *lip* has two small cartilages. The *teeth*, in both jaws, are granular, and crowded together. On the fore part of the palate is a kidney-shaped bone, covered with teeth like those in the jaws. The *tongue* is blunt, tied down, and smooth. The foremost aperture of each *nostril* is close to the jaw, and circular; and the hinder one is higher up, and oblong. The *eyes* are placed in the middle, are large, and look downward; and the longest axis of their pupil is vertical. Each *gill-cover* contains three plates, and leaves exposed its membrane, which contains eight rays.

The *back* slopes very little, and is not so protuberant as the *belly;* before the fin it has a sharp ridge. The hinder part of the belly, and all the lower side of the tail, have also sharp edges. The end of the *tail* is rounded, and a little dilated. The *lateral line* is above the middle, and is straight and smooth. The *vent* is near the middle.

The *back* fin slopes very suddenly from the first ray, which is a prickle, indented behind, and terminated by a substance like whalebone; there are, besides, seven branched rays. The *pectoral* fins are low, sharp above, and of moderate length: each contains twelve rays, of which the first is a prickle, indented on both edges; the others are branched. The *ventral* fins are very small, and at a distance from each other: each contains six soft rays, of which the first is undivided. The fin behind the *vent* does not approach that of the tail, and, near the body, is covered by a thick skin. It contains twenty-nine soft rays, of which the first four are undivided and contiguous to each other, and increase much in length from the first, which is very short. The other twenty-five rays decrease gradually in length to the end, which is rather angular, and are branched. The *tail* fin is long, and deeply divided into two sharp lobes. It contains seventeen distinct rays, with some short ones at each side.

V. Genus.—Malapterurus.

Fishes of the fifth order, with a smooth slimy skin, destitute of scales, with the mouth surrounded by tendrils, and with one fin on the back, placed toward the tail, and destitute of rays.

1st Species.—MALAPTERURUS COILA.

A *malapterurus* with eight tendrils.

This fish is found in the fresh water rivers of Bengal, and grows to about a span in length. It is not an ugly animal, and, by the natives, is considered good eating. It is *shaped* somewhat like the blade of a knife, being long, tapering, much compressed at the sides, blunt on the back, and sharp on the under edge. The whole almost is of a silver *colour*, with some black on the back and on the fin of the tail.

The *head* is oval, small, but wider than the body, and blunt. There are eight *tendrils*, slender, nearly equal in length, and reaching to about the middle of the fish. The *mouth* is small, and extends straight back, the under *jaw* being the shortest. The *lips* are fleshy. The *nostrils* are circular, and near the jaw. The *eyes* are of a moderate size, near the middle, round, and protuberant. The *gill-covers* expose part of their *membranes*; but I could not exactly ascertain the number of rays which each contains. Under the lower jaw there is no *slit*.

The *back* forms a smaller arch than the *belly*. The *tail* is very long, and its end is bent down. The *vent* is far forward. On each side are three smooth *lateral lines*.

The *back* fin is small and oval, with its widest end uppermost. The *pectoral* fins are low, and sharp above, and extend beyond the vent: each contains sixteen rays, of which the first is undivided, flexible, and smooth-edged; the others are branched. The *ventral* fins are minute, each containing seven rays, of which one on each side is undivided. The fin behind the *vent* contains seventy-one rays of nearly equal length, the first, second, and third, being undivided, and the others bifid.

The *tail* fin is divided into two sharp lobes, of which the under is the longest, and contains sixteen distinct rays, besides many small compacted ones.

VI. Genus.—Ageneiosus.

Fishes of the fifth order, with a smooth slimy skin, destitute of scales; with the mouth devoid of tendrils; and with two fins on the back, the hinder one having no rays.

1st Species.—Ageneiosus mino.

An *Ageneiosus* with eight rays in the back fin, and twenty-one in that behind the vent.

This fish is found in the upper part of the estuaries, that is, where the water possesses little or no saltness. It grows to about a foot and a half in length, and by the natives is considered good to eat. Its lower part is silver *coloured;* its upper parts green, changing to purple; but all its colours are lurid. In proportion to its breadth it is very long, and is but little compressed on the sides.

The *head* is of a moderate size, flattened above, blunt, and smooth; but the bone on the upper part is almost bare, and on its middle is a long deep furrow. On the nape is a bone marked with lines. The under *jaw* is the longest. The *mouth* is very large, and extends straight back from the extremity of the nose. The *lips* are fleshy, and, towards the corners, are reflected. The *teeth* in both jaws, and in a belt parallel to each side of the upper one, are crowded, conical, and blunt. Both apertures of the *nostrils* are circular. The *eyes* are far forward on the sides of the head, and their longest axis is horizontal, while that of their pupils is vertical. The *covers* of the gills are rounded behind, and do not conceal their *membranes*, each of which contains five rays. There is no *slit* under the lower jaw.

The *back*, from the fin to the head, descends much, and is covered with a bony plate marked with lines, and extending to the nape. At each side it is divided into two processes, the foremost of which descends to the opening of the gills. The *shoulder-bones* are almost naked, and end in a sharp point. The *lateral lines* are bent downward. The *vent* has two openings, and is behind the middle.

The first fin of the *back* is before the middle, and has eight rays, of which the first is a very short thick bone; the second is a very strong prickle, indented with blunt teeth before, and with sharp ones behind; the other six rays are branched, and the last of them is divided to the bottom into two. The second back fin is far behind the middle. The *pectoral* fins are small, and each contains nine rays, of which the first is like the second of the back fin. The rays in each *ventral* fin are six in number, and of the fin behind the *vent* twenty-one, of which the first six are undivided, and gradually lengthen from the first; the last fifteen are branched. The *tail* fin is divided into two sharp lobes, containing nineteen distinct rays, with some small compacted ones at the sides.

VII. Genus.—Pimelodus.

Fishes of the fifth order, with a slimy skin, destitute of scales or fleshy warts; with the mouth surrounded by tendrils; and with two back fins, of which the hinder one has no rays.

1st Species.—Pimelodus silondia. Plate VII. Fig. 50.

A *Pimelodus* with two tendrils, and an opaque body.

This fish is very common in the Gangetic estuaries, and is considered by the natives as good for eating. It commonly grows to three feet in length, and occasionally to twice that size. In *proportion* to its breadth it is very long, and is much compressed at the sides, with the upper edge rounded, and a great part of the lower edge sharp, somewhat like a knife. The back is of a dusky green *colour;* and, although the sides are like silver, the fish has a dirty lurid appearance, with a shade of livid hue. The back and tail fins are greenish, the others are white.

The *head* is short, thick, but not flattened, oval, with little slope, smooth, and covered with skin. On its summit is a longitudinal furrow. A *tendril* proceeds from each side of the upper jaw, but scarcely reaches farther than the corner of the mouth. The *mouth* is of a moderate size; and, when shut, descends backward. The under *jaw* is rather the longest. The *lips* are fleshy. The *teeth* in each jaw, and on a crescent-shaped bone on the fore part of the palate, are sharp, of unequal size, numerous, crowded, and of moderate size: on the four bones at the throat, they are very small and imbricated. The *tongue* is tied down, smooth, and blunt. The *nostrils* are small and circular; the foremost aperture is close to the jaw, and the other near it. The *eyes* are far forward, high, and convex; the greatest axis of their pupil is vertical. Each *gill-cover* is rounded behind, and exposes to view its *membrane*, which has ten rays. Under the throat is a *slit*.

On the *body* there are no naked bones. The *back* descends very little, and projects less than the *belly*, which is blunt-edged to the ventral fins. The *tail* diminishes gradually, and is sharp-edged below. The *lateral line* descends in a curve, with its concavity towards its back. The *vent* is rather before the middle.

The first fin of the *back* is far forward, and diminishes rapidly backward from the second ray, which is the longest; the first is a strong compressed prickle, with blunt knobs before, and sharp teeth behind; there are besides seven branched rays; but the two last approach near each other at their root. The hinder fin is far back, and oval, with its widest end most remote from the body. The *pectoral* fins are low, sharp above, and rather shorter than the head; each contains fourteen rays, of which the first is a prickle like that on the back. The *ventral* fins are still shorter than the pectorals, and are remote from each other; each has six rays, of which the first is undivided. The fin behind the *vent* ends at the back in a sharp point, and slopes very gradually from the fifth ray, which is the longest, those before it shortening very quickly, and like it being undivided, while the other thirty-six are branched. The fin of the *tail* is somewhat triangular, with the end like a crescent. It contains seventeen distinct rays, besides several short compacted ones on each side.

2d Species.—Pimelodus chandramara.

A *Pimelodus* with two tendrils, and a diaphanous body.

This little fish, which seldom exceeds an inch and a half in length, I found in the Atreyi river. It has some affinity to the *Pimelodus guttatus* of La Cepede, (*Hist. des Poissons*, Tome V. p. 96.) It is of a long *form*, flattened at the sides, and has a prominent belly. Except where the viscera are enclosed by a shining silver *coloured* membrane, and along the spine, which is blackish, it is every where diaphanous, here and there marked with clusters of black dots; and on each side is a golden stripe accompanying the lateral line.

The *head* is blunt, of a moderate size, and smooth. The *tendrils* are so minute as to be scarcely visible without a lens; but one proceeds from each corner of the mouth, and is shorter than the head. The *eyes* are large and protuberant. The *gill-covers* leave exposed their *membranes*, which contain rays, but, on account of their slenderness, their number cannot be ascertained. There is no *slit* under the throat. The *nostrils* and *teeth* are too minute for observation.

The *shoulder-blades* are naked.

The foremost *back* fin contains eight rays, of which the first is a smooth-edged prickle. The hinder fin is dotted. The prickle of each *pectoral* fin is indented behind. Each *ventral* has six rays, and the *vent* fin has seventeen. The fin of the *tail* is divided into two lobes.

3d Species.—PIMELODUS PANGASIUS. Plate XXXIII. Fig. 52.

A *Pimelodus* with four tendrils; with the tail fin bifid; and with the prickles of the back fin and pectorals very little longer than the membrane.

This fish is common in the estuaries of Bengal, and often grows to three feet in length. It is eaten by all ranks of natives that use fish, but is not considered as of the best quality. It is of a lurid appearance, long in *proportion* to its breadth, and flattened at the sides, and slopes gradually from the beginning of the foremost back fin toward both extremities. The *colour* above is dusky green, and below white, with a purple gloss on the sides. The fins are whitish.

The *head* is oval, short, rather wide, a little depressed, and blunt, with a notch on each side of the nose. On each side of its summit is a bony ridge slightly tuberculated. Two of the knobs on each are conspicuous, one at the fore end, and the other opposite to the eye. Two of the *tendrils*, from the middle of the upper jaw, are nearly as long as the head; and two still shorter proceed from under the chin. The *mouth* is moderate, and below the extremity of the nose, and extends straight back. The upper *jaw* is the longest. The *lips* cover the jaws. The *teeth*, both on the jaws and palate, are sharp, equal in size, very small, and crowded together. The fore part of the palate has a crescent-shaped bone covered with teeth, and divided into four lobes. At the throat there are four bones covered with teeth; the two above are half spheres; the under ones diverge, are narrow, and end in a point. The openings of the *nostrils* are circular; the foremost at the notch above the jaw, and the other nearer the jaw than the eye. The *eyes* are half way up the head, far forward, flat, and of a moderate size. The *gill-covers* have one plate with little covering, and are marked with waving lines, end in a point, and expose to view a thick *membrane*, containing ten rays on each side. Under the throat there is no *slit*.

The ridge of the *back* before the fin is rather sharp, and is covered by a waved bone, which, on each side, sends down a process to join the *shoulder-bone*, and this projects with a point over the pectoral fins. The *sides* slope outward. The *lateral line* is high up, straight, and smooth; and the *skin*, although opaque, is marked by the muscles, the marks uniting at the lateral line. The *vent* is nearly in the middle. The *tail*, near its end, is a little dilated, and divided into two lobes.

The foremost *back* fin is near the head, and slopes rapidly from the end of the second prickle to that of the last ray. It contains nine rays, of which the first is a very short triangular bone, the second a long strong prickle, ending in a substance like whalebone, and being indented on both edges; the other seven rays are branched. The hinder fin is far back, and somewhat of a hooked form, but blunt. The *pectorals* are low, nearly as long as the head, and sharp above; each contains twelve rays, of which the first is like the second in the back fin; the others are branched. The *ventral* fins are much smaller than the pectorals, and are separated some distance from each other: each has six soft rays, of which the first is undivided. The fin behind the *vent* ends sharp behind, and contains thirty-one rays, of which the foremost five are undivided, and rapidly increase in length from the first, which is very short: the others are branched, and slowly diminish in length to the last, which is deeply divided into two portions. The *tail* fin is divided into two sharp lobes, containing seventeen distinct rays, besides several short compacted ones at each side.

4th Species.—PIMELODUS RITA. Plate XXIV. Fig. 53.

A *Pimelodus* with six tendrils shorter than the head; with the tail fin two-lobed; with eight rays in the foremost fin of the back, the great prickle of which is smooth-edged; with twelve rays in the fin behind the vent; and with an opaque and unspotted body.

This fish is common in the estuaries of Bengal, grows to three or four feet in length, and is a clumsy lurid animal, which many of the Hindus will not eat. It is of a long conical *shape*, very little compressed, and tapers gradually toward both ends from the commencement of the foremost back fin. The upper parts are of a dusky *colour*, witha greenish gloss, changing to gold and purple, but of very dirty shades. The under parts are whitish. The fins are stained with red. The eyes are yellow.

The *head* is short, wide, depressed, oval, blunt, and covered with a slimy skin. Two very short *tendrils*, compressed at the root, proceed from before the hinder openings of the nostrils; two, a little shorter than the head, and having at their root a short membrane, proceed from the sides of the upper jaw; and two, of nearly the same length, proceed from the throat. The *mouth*, placed below the extremity of the nose, is of moderate size, and extends straight backward. The *lips* are fleshy. The *teeth* in both jaws, and on the palate, are granular, blunt, and crowded. In the jaws the anterior ones are the largest. On the fore part of the palate are two large oblong bones, covered with teeth. Both apertures of each *nostril* are circular: the one close to the jaw is large, and surrounded by a margin; the other, half way between the eye and jaw, is smaller. The *eyes* are above the middle of the cheek, and forward; both orbit and pupil are oblong, with their longest axis horizontal. Each *gill-cover* consists of one small smooth plate, leaving exposed a thick membrane, which contains many rays very indistinctly seen. There is no *slit* under the throat.

The ridge of the *back* is blunt, and is covered by two rough bones; the foremost is large and oblong, with three blunt processes toward the head, the largest in the middle; the other bone, adjacent to the fin, is heart-shaped. On each shoulder, also, is a rough bone of an oval shape, from which the *lateral line* descends straight to the end of the *tail*, which is dilated and rounded at the end, where it is slightly compressed. The *shoulder-blades* are naked, and very long, but blunt. The *vent* is behind the middle.

The anterior fin of the *back* is higher than the body, and slopes very suddenly backward. The foremost of its rays is a very short rough triangular bone; the second is an enormous smooth-edged prickle, tapering to a sharp point, and without any tip of the substance like whalebone; the other six rays are branched; and the last is divided to the root. The hinder fin is almost as long as that behind the vent, to which it is opposite, rises very gradually, and toward the tail ends abruptly. The *pectoral* fins are low, about as long as the head, and sharp above: each has twelve rays, of which the first is a very strong flat prickle, indented on both edges, and terminating in a substance like whalebone; the teeth on its fore edge are turned toward the point, those behind toward the root. The *ventral* fins are smaller than the pectorals, and each contains eight rays, of which the first and last are undivided. The fin behind the *vent* is somewhat rounded, but slopes more behind than before: its first four rays are undivided, and the others are branched, the last being divided to the very root. The fin of the *tail* consists of two unequal lobes, the upper being the longest and sharpest. It has fifteen distinct rays, besides some short compacted ones at each side.

5th Species.—Pimelodus gagora. Plate X. Fig. 54.

A *Pimelodus* with six tendrils shorter than the head; with the tail fin divided into two lobes; with nine rays in the first fin of the back, which has no elongation beyond its membrane; but its prickle is indented on both edges; with eighteen rays in the fin behind the vent; and with the sides unspotted, and opaque.

The *Gagora* has a considerable affinity to the *Pimelodus barbus* of La Cepede, (*Hist. des Poissons,* Tome V. p. 106,) a fish of South America, and therefore probably different from that of India. It is common in the estuaries of Bengal, grows to about three feet in length, and is considered good eating by the natives. The *form* is long, conical, little compressed at the sides, and rather more prominent above than below. From the beginning of the back fin it slopes gradually to both ends, but below the tail diminishes rather suddenly at the fin behind the vent. The *colour* above is a dusky purple, and below white. The sides have a gloss like silver. The back fins are dotted with black, the dots on the hinder fin forming a faint large spot. The fin behind the vent, that of the tail, and the ventrals, are white, stained with red.

The *head* is short, but thick, oval, lightly depressed, and rises with a convex slope from the jaw to the nape; and this slope continues to the back fin. From this organ there extends forward a large bony plate, which covers the fore part of the back and the hind part of the head, and is marked with waved lines. On the nape, as from a centre, there diverge, in different directions, several straight lines. On the centre of the fore part of this plate is an oblong cavity, sharp at each end; and behind the eye the plate forms into four separate ridges, one of which, on each side, passes to the nostril, and the other two re-unite on the nose. Behind the plate has also four short processes, one on each side, running down the shoulder, and another on each side of the first ray of the back fin. The two *tendrils*, from the sides of the upper jaw, are almost as long as the head, and the four from the chin are shorter. The *mouth* is of moderate size, placed under the extremity of the nose, and extending straight backward. The under *jaw* is the shortest, and in its middle has a slight longitudinal ridge. In each jaw are many granular *teeth*. On the palate are two bones covered with sharp crowded teeth. The apertures of the *nostrils* are near each other, and about half way between the eyes and jaw. The foremost aperture is circular; the hinder crescent-shaped, and covered by a reflected membrane. The *eyes* are of moderate size, high, and forward; and the pupils are oblong. Each *gill-cover* consists of one plate, which exposes to view a *membrane* containing four flattened bones. Under the chin is a *slit*.

The *back* forms a sharp ridge. The *sides* slope outward, and on each shoulder, above the blade, which is naked, there is an oblong protuberance. The *lateral line* descends from the upper side of this protuberance, is smooth, and very distinctly marked. The *vent* is behind the middle. The *tail* is dilated at the extremity, and ends in a notch.

The first fin of the *back* is far forward, and slopes very rapidly from the point of the second ray to the end. The first of its rays is a very short bone; the second is a strong prickle, terminating in a substance like whalebone, and indented on both edges, and the others are branched. The second fin is almost as long as that of the vent, and rises gradually to the end, where it is rounded above, but sharp below. The *pectoral* fins are shorter than the head, and low: each contains twelve rays, of which the first is a prickle like that of the back; the others are branched. The *ventral* fins are shorter than the pectorals, and near each other: each contains six rays, of which the first is undivided. The fin behind the *vent* is nearly triangular, sloping much, both before and behind, from the sixth ray, which is the longest; the first five are undivided. The fin of the *tail* is divided into two lobes, of which the upper is rather the longest and sharpest: it contains eighteen distinct rays, besides some short ones on each side.

6th Species.—PIMELODUS SAGOR.

A *Pimelodus* with six tendrils, rather shorter than half the fish, with the fin of the tail divided into two lobes, with nine rays in the back fin, which has no elongation beyond its membrane, but its prickle indented on both edges; with eighteen rays in the fin behind the vent; and with opaque sides transversely barred above the lateral line, which is single.

The *Diddi Jellah* of Dr Russell (*Indian Fishes*, Vol. II. No. 169) is very like the *Sagor*, but has two lateral lines on each side. The *Silurus Herzbergii* of Bloch (*Ichth.* Tome XI. p. 24, Pl. CCCLXVII.) has also a considerable affinity to the *Sagor*, but differs in colours, being brown above and silver below, and also in having only thirteen rays in the fin behind the vent.

The *Sagor* is found along with the *Gagora*, is nearly of the same size, and the two differ more in appearance than in structure. So like in this respect are they, that here I shall only mention the circumstances in which the *Sagor* differs from the description given of the *Gagora.*

The *Sagor* above is of a green *colour*, with a golden gloss, and adorned with several golden belts, descending to the lateral line. The naked bone covering the *head* is distinct from that on the back, is rounded before, and terminates behind in four lobes. The bone on the *back* consists of two rounded lobes. On each *gill-cover* are two naked bones, roughened with tubercles; the foremost being long and sharp at each end. The two upper *tendrils* reach to the middle of the pectoral fins, and the two outer ones on the chin are only a little shorter. There is only one bone on the *palate*, and it is divided into four lobes. The *lateral line*, in its fore part, is covered with tubercles. The *pectoral* fins have only eleven rays each. The first seven rays of the *vent* fin are undivided. In the *tail* fin there are only sixteen distinct rays.

7th Species.—Pimelodus arius.

A *Pimelodus* with six tendrils not longer than the head; with the fin of the tail divided into two lobes; with eight rays in the back fin, the prickle being indented on both edges, and ending in a flexible point much longer than the membrane; with twenty-two rays in the fin behind the vent; with teeth on the palate; and with the sides unspotted and opaque.

This is another fish, so like the *Gagora*, that, besides the circumstances mentioned in the specific character, I shall only notice what it differs in from the description of that animal.

The *Arius* (*Ari gagora*) is found in the same places with the *Gagora*, but seldom exceeds two feet in length. Its *colours* are similar to those in the description; but the spot on the second back fin is more distinct, nor have any of the back fins a tinge of red. On its palate are two triangular bones, thinly covered with blunt *teeth.* The *membrane* of each gill-cover contains five rays. The first ray of the *back* fin is very short, and the last is divided to the very root. There are eleven rays in each *pectoral* fin; the first seven of those in the fin behind the vent are undivided. There are only fifteen distinct rays in the fin of the tail.

8th Species.—PIMELODUS JATIUS.

A *Pimelodus* with six tendrils not longer than the head; with the fin of the tail divided into two lobes; with eight rays in the back fin, the prickle being indented on both edges, and ending in a flexible point much longer than the membrane; with twenty-one rays in the fin behind the vent; with no teeth on the palate; and with opaque sides.

The *Jatius (Jat gagora)* is a third species, which differs so little from the *Gagora*, that I shall only mention the few particulars, besides those contained in the specific character, in which it differs from the description of that fish.

The *Jatius* is found in the same place, grows to the same *size*, and has similar *colours* with the *Gagora,* only the spot on the hinder back fin is more distinct. On the palate there are no bones covered with *teeth;* but there is a kind of membranous veil behind the upper *jaw.* There is no *slit* on the throat. The first fin of the *back,* the *pectorals,* and the fin behind the *vent,* entirely resemble those of the *Arius* just described, only the last contains one ray less.

8th Species.—PIMELODUS NENGA.

A *Pimelodus* with six tendrils longer than the head; with the tail fin divided into two lobes; with nine rays in the foremost back fin, the prickle being indented on both edges, and ending in a flexible point, much longer than the membrane; with twenty rays in the fin behind the vent; with teeth in the palate; and with opaque sides marked with short transverse lines.

The *Nenga* (*Nenga gagora*) is a fourth species equally allied to the last four as they are to each other, and found in the same places. Besides the circumstances mentioned in the specific character, it differs from the *Gagora* as follows:

The *colour* is golden, with a dusky shade on the back. All the fins are yellowish, stained with black. On each side, above the *lateral lines*, are some short vertical marks. The bones, covered with *teeth*, are situated on the middle of the palate, close to each other. The *tendrils* extend beyond the pectoral fins. The bony plate on the head and shoulders wants the ridges that go to the nostrils, and its hinder branches are shorter. The foremost seven rays of the fin behind the *vent* are undivided.

9th Species.—PIMELODUS SONA.

A *Pimelodus* with six tendrils longer than the head; with the tail fin divided into two lobes; with nine rays in the first fin of the back, the prickle being indented on both sides, and ending in a flexible point much longer than the membrane; with eighteen rays in the fin behind the vent; and with opaque spotless sides.

This is still another kind found in the same places, and so nearly allied to the above, that, besides the differences mentioned in the specific character, I shall only notice the circumstances in which it differs from the description of the *Gagora*.

The gloss on its sides is golden. The two upper *tendrils*, and the two outer of those below the chin, reach to the end of the pectoral fins. The *teeth* are sharp. The *eyes* are oblong, with the longest diameter horizontal, while the longest diameter of the pupil is vertical.

10th Species.—Pimelodus viridescens. Plate XI. Fig. 56.

A *Pimelodus* with six tendrils; with the tail fin divided into two lobes; with seven rays in the foremost fin of the back, the prickle having both edges smooth; with thirteen rays in the fin behind the vent; with the back reddish-brown, crossed by three green bars; and with opaque spotless sides.

This small fish is found in the rivers of the northern parts of Bengal. It is depressed in the fore part, and conical behind. The sides are silver *coloured*, the belly livid, and the fins of the back and tail spotted.

The *head* is very large, depressed, and blunt. On its summit are two longitudinal bony ridges, with a cavity between. The two *tendrils*, from the sides of the upper jaw, are about as long as the head; and the four from the chin are somewhat shorter. The *mouth*, situated a little below and behind the extremity of the nose, is large. The jaws are covered by *lips*. There are no *teeth*, except a slight roughness on the jaws. The apertures of each *nostril* are close to each other, and half way between the eyes and jaw. The *eyes* are high, of moderate size, and oblong. The *gill-covers* expose their *membranes*, each of which contains six very slender rays. Under the throat is a *slit*.

On the fore part of the *back* there are two rough bony plates, the foremost of which is the largest, and behind very deeply divided into three lobes. Each *shoulder-bone* sends a double naked process above the pectoral fin. The *lateral line* is straight, and high on the side. The *vent* is behind the middle.

The foremost fin of the *back* contains seven rays, of which the first is a very short thick bone; the second is a strong sharp prickle, obliquely striated, but not indented on either edge; the other five are branched, the last being divided to the root. The hinder fin is nearly triangular. The *pectoral* fins are low, and each contains about nine rays, of which the first is a very strong prickle, indented behind. The *ventral* fins are shorter than the pectorals, and near each other; and each contains six rays. The fin behind the *vent* contains thirteen rays: the three first are simple, and increase gradually in length from the first, which is very short. The *tail* fin is divided into two lobes nearly equal: it contains about twenty distinct rays.

11th Species.—Pimelodus cenia. Plate XXXI. Fig. 57.

A *Pimelodus* with six tendrils not longer than the head; with the tail fin divided into two lobes; with eight rays in the back fin, the prickle in which has smooth edges; with thirteen rays in the fin behind the vent; with the sides silvery; and with the back marked with four black transverse bars.

This fish, which is about three inches in length, is found in the northern parts of Bengal, where it frequents rivers. It is long in proportion to its breadth, of a compressed *form*, and slopes with a gentle curve, above and below, from the commencement of the first back fin to both extremities. The general *colour* is silver, with some dusky on the back, and the bars descending to the lateral lines. The end of the tail is black; and the first fin of the back, and that of the tail, are stained with the same colour. The eyes are silver-coloured.

The *head* is much shorter, and no wider, than the body, and slopes very suddenly from the nostrils to the extremity. On the nape is a bony plate, which sends seven ridges diverging from thence as a centre: the two foremost run forward near each other between the eyes; two short ones run toward each side; and the seventh runs to the back fin. The two *tendrils* proceeding from the upper jaw are as long as the head, flattened, and joined to the corner of the mouth by a short membrane: the four proceeding from the chin are very short. The *mouth* is below and behind the extremity of the nose, and is small; and, when it opens, the jaws protrude more than usual in fishes allied to this, but only a very little. The middle of the upper *jaw* has a longitudinal ridge, which is received in a notch of the under one. The *lips* are fleshy. The *teeth* are scarcely perceptible, and each *nostril* has its apertures near each other, and in the middle between the eyes and jaw. Both are circular, and the pupils of the eyes are oblong. The *eyes* are high and forward, and are of moderate size. The *membrane* of each gill-cover is exposed, and contains five rays. There is no slit under the throat.

The *back* is rounded across, and, before the fin, has a small triangular bony plate. The *lateral line* runs straight along the middle of the side, which is marked by the separations of the muscles. The *vent* is near the middle.

The first fin of the *back* is far forward, and slopes much behind. The first of its rays is a very short bone; the second is a strong prickle, striated on the sides, but smooth on the edges; and the others are branched. The second fin is very low, and nearly triangular. The *pectoral* fins are very low, as long as the head, and sharp above; each contains nine rays, of which the first is a strong prickle, streaked on the sides, and indented behind. The *ventral* fins are much smaller than the pectorals, and remote from each other; each contains six rays, of which the first is undivided. The fin behind the *vent* slopes much backward. The three first of its rays are undivided, closely joined, and lengthen gradually from the first, which is very short; the ten last are branched, and the last of them is split to the bottom. The *tail* fin is divided into two very sharp equal lobes.

12th Species.—Pimelodus rama. Plate III. Fig. 55.

A *Pimelodus* with six tendrils shorter than the head; with the fin of the tail divided into two lobes; and with a diaphanous body.

This small fish, which I found in the Brahmaputra river, so strongly resembles the *Chandramara*, already described, (page 162,) that I shall only mention those circumstances in which it differs from the description of that species.

The *colour* is yellowish, without dots. On the nape is a large black spot, divided into four lobes. The six *tendrils* are distinctly visible, and are thicker than the two which the *Chandramara* has. In the fin behind the *vent* there are fifteen rays.

13th Species.—Pimelodus tengana. Plate XXXIX. Fig. 58.

A *Pimelodus* with eight tendrils all shorter than the head; with the tail fin bifid; with a diaphanous body dotted on the upper parts; and with fourteen rays in the fin behind the vent, and eight in the foremost on the back.

This is a pretty little fish, which grows to about three inches in length, and is found in the Brahmaputra river. It is long in *proportion* to its breadth, more prominent on the back than below, and compressed on the sides; and tapers gradually to both ends from the commencement of the first fin on the back. The body is diaphanous, with a silver *coloured* membrane investing the viscera and spine, and with a gloss of gold on the sides. On the back are many black dots, which are collected into a spot above each pectoral fin, and also on the crown of the head. The fins of the back and tail also are dotted, so that the edge of the last is black, and several spots are formed on the first.

The *head* is short, rather thick, oval, and a little sharp, with a slight ridge above each eye. The *mouth*, extending straight back, is small, and beneath and behind the extremity of the nose. The *eyes* are high, and of moderate size. Both openings of each *nostril* are circular, with a tendril between them. The *membrane* of each gill-cover contains six rays. Under the lower jaw there is no *slit.*

The *back* slopes considerably from the first fin to the end of the nose, and forms a sharpish ridge, covered with a narrow bone, which is divided behind, to surround the first ray of the fin. The *shoulder-bones* are naked, and the *lateral lines* straight. The *vent* is near the middle, and the *belly* is a little prominent, and convex across.

The foremost *dorsal* fin slopes much backward. The first of its rays is a very short bone; and the second is a strong prickle, with smooth edges. The second fin is much shorter than the fin behind the vent, and somewhat rounded. The *pectoral* fins are shorter than the head, low, and sharp above: each, besides soft rays, has a strong prickle, indented behind. The *ventral* fins are still smaller than the pectorals, and each contains six rays. The fin behind the *vent* slopes backward, and contains fourteen rays, of which the three foremost are undivided. The *tail* fin is divided into two equal sharp lobes.

14th Species.—PIMELODUS URUA.

A *Pimelodus* with eight tendrils longer than the head; with the tail fin bifid; with a diaphanous body dotted above; and with forty-three rays in the fin behind the vent, and five in the foremost on the back.

This is a little fish, very similar to that last described, and it seems to have some affinity to the *Pimelodus atherinoides* of La Cepede, (*Hist. des Poissons*, Tome V. p. 96.) I found it in the rivers and ponds of the northern parts of Bengal. It grows to about three inches in length, and is diaphanous, with black dots on the upper parts of the body and head, while a silver *coloured* membrane invests the viscera and spine.

The *head* is very short, but wider than the body, and rather flat. On the nape is a naked dotted plate, which sends forward two ridges, terminating at the end of the nose, and, backward, a third ridge, which extends to the commencement of the foremost dorsal fin. The two *tendrils* between the nostrils, and the four from under the chin, are a little longer than the head; while the two from the sides of the upper jaw are longer than the pectoral fins. The *mouth* is below the end of the nose, extends straight back, and is of moderate size. The *jaws*, in place of teeth, have a roughness on their edges. The *eyes* are globular, high, and of moderate size. The *gill-covers* end in a point, and expose their *membranes*, each of which contains about nine very slender rays. There is no *slit* under the throat.

The fore part of the *back* slopes toward the head. The *belly* is prominent. The *vent* is before the middle. The *lateral lines* are faint, and run straight along the centre.

The foremost fin of the *back* has five or six rays, of which the first is a prickle, indented behind: the hinder fin is very small. The *pectoral* fins are low, and shorter than the head, and each contains about ten rays, of which the first is a very strong prickle, barbed on the hinder edge. Each of the *ventral* fins contains about seven rays; but so slender, that I cannot be certain of the number. The same is the case with those of the fin behind the *vent*, which are about forty-three in number. The fin of the *tail* is divided into two equal sharp lobes, and contains about twenty-four rays.

15th Species.—Pimelodus batasio. Plate XXIII. Fig. 60.

A *Pimelodus* with the fin of the tail bifid; with eight tendrils shorter than the head; with a diaphanous body, having on each side two stripes dotted with black; with sixteen rays in the fin behind the vent; and with nine in that of the back.

This small fish I found in the river Tista. It grows to about three inches in length, and its skin is smooth. The cavity containing the viscera is lined with a silver *coloured* membrane; and, besides the two stripes, the fish has on each shoulder a spot composed of black dots. The first back fin is spotted, the second is dotted.

The *head* is oval, small, and covered with skin, and along its summit there runs a cavity. There are eight *tendrils*, much shorter than the head. The mouth is small, and extends directly backward from below, and from a little behind the extremity of the nose. The *lips* are fleshy. The *teeth* in both jaws are minute and crowded. The *nostrils* are in the middle between the jaw and *eyes*, which are small and high. The *gill-covers* expose part of their *membranes*, which are diaphanous, and each contains about six rays. There is no *slit* under the throat.

The *back* slopes down from the fin to the head, forms a sharp ridge, and, on its fore part, is covered by a bony plate, shaped like a heart, and marked by a longitudinal cavity. On each *side*, above the pectoral fin, is an oblong tumour. The *shoulder-bones* are naked, and end in a sharp point. The *lateral lines* run straight along the middle of each side. The *belly* is straight, and transversely obtuse. The *vent* is near the middle of the fish, and a furrow runs from it to the commencement of the fin.

The first ray of the foremost fin of the *back* is a very short bone; the second is a strong prickle, smooth on both edges, and the others are branched. The hinder fin is oblong. Each *pectoral* fin contains a very strong prickle, indented behind. Each *ventral* fin has six rays. The three first rays of the fin behind the *vent* are undivided, and lengthen gradually from the first, which is very short. The *tail* fin is divided into two equal lobes, and contains about twenty rays.

16th Species.—PIMELODUS ANGIUS. Plate XXIX. Fig. 59.

A *Pimelodus* with the fin of the tail bifid; with eight tendrils longer than the head; with the body somewhat diaphanous, and having on each side three dotted longitudinal stripes; with forty-two rays in the fin behind the vent; and with seven in that of the back.

This fish is found in the rivers of Bengal, is very beautiful, and grows to four or five inches in length. It is long in *proportion* to its breadth, and very much flattened on the sides. The belly is more prominent than the tail, which is rather bent downward at the end, so that the fish has a strong affinity in structure to the *Silurus Pabo.* The body is smooth, and somewhat diaphanous, and has a bright gloss, changing from gold to purple. A silvered *coloured* membrane lines the visceral cavity. There are seven longitudinal stripes, composed of black dots, one on the back, and three on each side: of these the undermost has two branches at its fore end. Along the lateral line is a pale golden stripe, ending at the tail in a round black spot.

The *head* is very small, oval, and little compressed, and along the crown has two ridges. The two *tendrils* from between the nostrils, and the four from under the chin, are a little longer than the head; the two from the corners of the mouth extend to the ventral fins. The *mouth* is small, and below the extremity of the nose, and extends straight back: the upper *jaw* is the longest. The *nostrils* are near the jaw, with the foremost aperture circular, and the hinder one oblong. The *eyes* are large, and circular; but the pupils have their longest diameter placed vertically. The *gill-covers* are rounded behind, move freely, and contain one smooth bony plate. Their *membranes* are exposed to view, and contain rays, the number of which I could not exactly ascertain. There is no *slit* under the throat.

The *back*, forming a ridge, slopes from the first fin to the head. The *belly* is prominent, and rounded. The *tail* tapers gradually, and below is sharp-edged. The *lateral line* is high, straight, and smooth. The *vent* is before the middle.

The first fin of the *back* slopes very rapidly backward, so that its edge is almost vertical. Its first ray is a strong smooth prickle, and the others are branched. The second fin is higher than its length along the back, and it widens towards the point, which is reflected. The *pectoral* fins are sharp above, and about as long as the head: each has seven rays, of which the first is a strong prickle, barbed on the hinder edge; the others are branched. The *ventral* fins are much smaller than the pectorals, and stand near each other: each contains six rays, of which the first is short, and undivided. The edge of the fin behind the *vent* forms a slight arch. It contains about forty-two rays, of which the first three are undivided. The fin of the *tail* is divided into two divaricating lobes, nearly equal in length, and convex on their posterior edge.

17th Species.—Pimelodus carcio.

A *Pimelodus* with the tail fin bifid; with eight tendrils, of which two only are longer than the head; with the body nearly opaque; with four dotted longitudinal stripes on each side; and with fourteen rays in the fin behind the vent, and eight in that of the back.

This fish I found in ponds in the northern parts of Bengal.

It is about three inches in length, long in proportion to its breadth, smooth, flattened on the sides into the *shape* of a wedge, and tapers rather suddenly at the second fin of the back, and at the fin behind the vent. The sides are transversely marked by the muscles. The *colour* is yellowish, with a gloss of silver, which predominates on a stripe along the lateral line. The four stripes on each side consist of black dots crowded together; and above the shoulder-bone is a large dark spot. The belly is white; and the fins on the back, and behind the vent, are dotted. The eyes are silver-coloured.

The *head* is short, wider than the body, oval, blunt, and not compressed. On its crown is a rough bare bone, which sends an angle down to the beginning of the lateral line, and a narrow process to the first fin on the back. Along its middle, between the eyes, is a long cavity. The two *tendrils*, from the corners of the mouth, extend beyond the end of the pectorals, but not to the end of the first fin on the back; the six others are shorter than the head. The *mouth* is small, and extends straight back, the *jaws* being nearly equal in length. The *lips* are fleshy; the *teeth* crowded and minute. One aperture of each *nostril* is close to the jaw, and the other half way to the eye. The *eyes* are high and small, and have circular pupils. The *gill-covers* are smooth, consist of one plate, and expose their membranes, each of which contains about six rays. There is no slit under the throat.

The *back* is rather more prominent than the *belly*, has a considerable slope from the fin to the head, and its ridge is formed by the process from the bony cover of that member, which terminates in a fork, surrounding the root of the prickle. The *lateral line* descends straight from the shoulder to the end of the tail. The *vent* is nearly in the middle.

The first fin of the *back* slopes considerably backward. Its first ray is a very short bone, the second a strong prickle, barbed behind, and the others are branched, the last being split to the root. The hinder fin is oblong, and rises gradually to near the end, which is very blunt. The *pectoral* fins are low, shorter than the head, and sharp above. Each contains about seven rays, of which the first is a very strong prickle, deeply indented behind. The *ventral* fins are still smaller than the pectorals, and each contains six rays. The fin behind the *vent* slopes backward. Its first four rays are undivided, and lengthen gradually from the foremost; the others are branched, and the last is divided to the bottom. The *tail* fin consists of two very sharp lobes, the posterior edges of which are convex.

18th Species.—PIMELODUS TENGARA. Plate III. Fig. 61.

A *Pimelodus* with the tail fin bifid; with eight tendrils, whereof four are longer than the head; with a smooth opaque body, having four dotted stripes on each side; with ten rays in the fin behind the vent; and with eight in that of the back.

The *Silurus vittatus* of Bloch (*Ichth.* Tome II. page 40, Pl. CCCLXXI. Fig. 2) has some affinity to the *Tengara*, but the prickle on the back is indented behind.

The *Tengara* is very common in the ponds of India, and by the natives is considered good eating. It is a beautiful little fish, from four to six inches in length. It is higher in *proportion* to its length than most species of *Pimelodus*, a good deal compressed at the sides, and its back is much more prominent than its belly. The upper ridge of the tail contracts rather suddenly at the end of the second fin. The upper parts are of a green *colour*, changing to purple, the under white. On each side are four longitudinal stripes, covered with black dots, and separated by the lateral line, and three pale yellow stripes. On each side above the pectoral fin is a large dark spot.

The *head* is very blunt, of moderate size, and half oval, sloping much more above than below. It is covered above by a bone, waved irregularly with many lines, and having a long hollow between the eyes. This bone terminates behind in three long processes; one in the centre, reaching up the back to the first fin, and one on each side, extending to the shoulder-bone, and ending in a point turned backward. The two *tendrils*, from between the nostrils, are much shorter than the head; the two intermediate ones from the chin extend to the end of the gill-covers; the two outer ones from under the chin extend a little farther; and the two from the upper jaw reach to the end of the tail. The *mouth* is below, and a little behind the end of the nose, is of a moderate size, and extends straight back. The upper jaw has a membranous veil behind the *teeth*, which, in each jaw, are minute and granular. The *tongue* is blunt, and tied down. The foremost aperture of each *nostril* is tubular, the other is nearly in the middle, between the jaw and eye. The *eyes* are small, round, and high. The *gill-covers* end in a point, are naked, and waved with lines, and each consists of one plate. Their *membranes* are exposed to view, and thick, each containing about six rays. There is a *slit* under the throat.

The *back* slopes, in the same line with the head, to the first fin; on each side of this is a small bony plate. The *shoulder-bone* is naked, and ends in a point. The *lateral line* runs straight along the middle of the side. The *vent* is behind the middle.

The foremost *dorsal* fin slopes gradually back, and its first ray is a short sharp prickle, smooth on both edges; the others are branched. The hinder fin is longer than that behind the vent, rises gradually backward, and terminates in a very blunt vertical end. The *pectoral* fins are low, shorter than the head, and sharp above. Each contains a strong prickle barbed behind, and seven soft rays. The *ventral* fins are still shorter than the pectorals, and each contains six rays, of which the first is undivided. The fin behind the *vent* slopes a little backward, and its three first rays are undivided, and lengthen gradually from the foremost, which is very short. The *tail* fin is divided into two equal sharp lobes.

19th Species.—PIMELODUS TELCHITTA.

A *Pimelodus* with the fin of the tail divided into two nearly equal lobes; with eight tendrils shorter than the head; with an opaque, rough, reddish body, having on each side one faint stripe; with thirteen rays in the fin behind the vent; and with eight in the first of the back.

This fish frequents the fresh water rivers of Bengal and Behar, grows to about a span in length, and, when taken out of water, is tenacious of life. It is long in *proportion* to its breadth, tapers gradually to the tail, and is scarcely compressed at the sides. The *colour* is reddish, with a golden gloss, having a brownish back, and, on each side, a longitudinal stripe of the same colour. The fins are diaphanous.

The *head* is small, oval, and sharpish. Above it is covered with bony plates, variously furrowed, of which the hindmost is divided into three parallel processes, extending almost to the first fin of the back. The two *tendrils* from the nostrils, and the four from under the chin, are very short. The two from the corners of the upper jaw, although longer than the others, are shorter than the head, and have a short membrane connected to their hinder edge. The *mouth*, placed under and behind the extremity of the nose, opens directly back. The *lips* are fleshy. There are no visible *teeth.* The apertures of each *nostril* are nearer the jaw than the eye, and are separated only by the tendril. The *eyes* are high and far back, and are minute with circular pupils. The *gill-covers* are rough, with small tubercles. Each *membrane* contains four or five very slender rays. There is no *slit* under the throat.

The *back* slopes very little toward the head. On the *sides* are scattered many small tubercles. The *lateral line* runs nearly straight along the middle of the side. The *vent* is rather behind the middle.

In the foremost fin of the *back,* the first ray is a very short bone; the second is a prickle smooth on the edges, and the others are branched. Each of the *pectoral* fins has ten rays, of which the first is a prickle indented behind. Each *ventral* fin contains six rays. The three first rays of the fin behind the *vent* are undivided, and lengthen gradually from the first, which is very short. The *tail* fin has nineteen distinct rays, besides some that are compacted.

20th Species.—PIMELODUS BAGARIUS. Plate VII. Fig. 62.

A *Pimelodus* with the tail fin divided into two very long-pointed unequal lobes; with eight tendrils not longer than the head; with a scabrous opaque body, variegated with black irregular marks; with thirteen rays in the fin behind the vent; and with eight in the foremost fin of the back.

The *Bagarius (vaghari)* grows often to a large size, and is sometimes above six feet in length. It is a very ugly animal, with lurid colours. The fish is very long in *proportion* to its breadth, and rather flattened, but more prominent above than below. It tapers gradually to the end of the tail from the first fin of the back. The skin above, although slimy, has many small tubercles, that give it a scabrous feel. The *colour* is greenish ash, very pale below; and not only on the body, but on the fins, are scattered large irregularly shaped black marks. The eyes are golden-coloured.

The *head* is very flat, large, and blunt. It is covered by a rough bone, which terminates behind in three narrow processes, whereof the middle one is the longest, and extends along the back to near the fin, while the two outer ones reach to the beginning of the lateral lines. The two *tendrils*, which separate the apertures of the nostrils, and the four from under the chin are very short; the two from the corners of the upper jaw are about as long as the head, and have behind a narrow membrane, extending to more than their middle. The *mouth*, placed below and behind the extremity of the nose, is large, and extends straight backward. The under *jaw* is the shortest, and has a longitudinal ridge on the centre. The *lips* are fleshy. In each jaw are crowded many *teeth* of unequal sizes, and some of them large. On the hinder part of the palate are two bones covered with teeth. There is no *tongue*. Both apertures of each *nostril* are large; the foremost, however, being smallest and lowest, and they are separated only by a tendril nearer the jaw than the eye. The *eyes* are very small, are far up and back, and have a pupil shaped like the ace of hearts. The *gill-covers* end in a long point, with a sinuosity below it, and consist of one scabrous plate. Each of their *membranes* contains eleven rays. There is no *slit* under the throat.

The *back*, before the fins, has a very slight declivity to the head. Before the first fin is a small rough bony plate, which sends a short branch along each side. The *lateral line* is straight, and descends from the shoulder to near the end of the tail, where it divides into two branches. The *vent* is behind the middle. The *tail* is a little compressed, and dilated towards the end, where it forms two small lobes.

The first fin of the *back* slopes backward, with a slightly concave edge. The first of its rays is a very short bone. The second is a prickle, smooth on both edges, and terminates in a long point, flexible like whalebone, which extends far beyond the membrane uniting the other six rays, which are branched. The second fin is rather shorter than that behind the vent, and is nearly triangular. The *pectoral* fins are low, sharp above, and longer than the head. Each has thirteen rays, of which the first is a strong prickle, smooth on both edges, and terminated by a flexible point, that extends far beyond the membrane joining the other rays, the two last of which are undivided. The *ventral* fins are much shorter than the head, and their edge forms a slight concavity. Each has six rays, of which the first is undivided. The three first rays of the fin behind the *vent* are undivided, and lengthen gradually from the first, which is short. The remaining ten are branched, and their ends form a concave edge. The *tail* fin is very long, and is divided into two lobes, of which the uppermost is the longest, and bent like a hook. Each ends in a long slender point; and, in the whole, there are seventeen distinct rays, besides some compacted ones on each side.

21st Species.—Pimelodus cavia.

A *Pimelodus* with the fin of the tail divided into two lobes, the undermost of which is the longest; with eight tendrils not longer than the head; with an opaque scabrous clouded body; with nine rays in the fin behind the vent; and with five in the foremost of the back fins.

This fish may possibly be the same with the *Pimelodus cous* of La Cepede, (*Hist. des Poissons,* Tome V. p. 110,) for the description given by him is so short, that one cannot be satisfied on the subject. What he states, however, respecting the colour of the *cous* is not very applicable to the *Cavia*, (*Kauya tenggara.*)

The *Cavia* is found in the northern rivers of Bengal, where it grows to about six inches in length, and retains life long after it is taken from the water. It is neither common, nor is it held in peculiar esteem for eating. It is flattened before, conical behind, and all its upper parts are scabrous from elevated spots. The *colours* are lurid, and the back is variegated with brown dots, collected into cloud-like marks, while the sides have a gloss like silver, changing into the hue of copper. The belly is of a dirty livid appearance. Across the tail are two transverse bars, and on the tail fin a third, all of which have fewer dots than the adjacent parts. The eyes are white.

The *head* is very large, a good deal flattened, blunt, and covered above with bony plates, forming various ridges. The two *tendrils* between the nostrils are very short; the two from the upper jaw are nearly of the same length with the head, the four under the chin are shorter. The *mouth*, placed below and considerably behind the extremity of the nose, is very large, and extends straight back. The *lips* are fleshy. Except a slight roughness on the jaws there are no *teeth*. The *nostrils* are nearer the jaw than the *eyes*, which approach each other, are small, and have oblong pupils. The *gill-covers* are rough, and expose their *membranes*, but these are so thick that the number of rays which they contain cannot be exactly ascertained; each, however, has about six. There is a *slit* under the throat.

Between the first fin and the head the *back* has two bony plates, of which the foremost is the largest, and behind is divided into two. The *shoulder-bones* are naked, and each sends a long process above the pectoral fin. The *lateral line* runs straight along the middle of the side. The *vent* is near the middle.

The first ray of the foremost fin of the *back* is a strong blunt prickle, rough, but not indented on the foreside. The hinder fin is small and scabrous. The *pectoral* fins are low, and each contains about seven rays, the first of which is a very strong prickle, indented on both sides. Each *ventral* fin contains six rays, and that of the *tail* about thirteen.

22d Species.—PIMELODUS HARA.

A *Pimelodus* with the end of the tail fin like a crescent; with eight tendrils, of which two are a little longer than the head; with an opaque scabrous brown clouded body, and some transparent spots on the fins; with ten rays in the fin behind the vent; and with seven in the foremost on the back.

This fish I found in the river Kosi. It is an ugly creature, of little or no value, and seldom exceeds three inches in length. The body is conical, of a lurid brown *colour*, with a copper gloss, and is clouded with several irregular black marks. The fins are of a similar colour, but have some parts transparent, somewhat similar to what occurs in the wings of certain moths. The four upper tendrils have numerous black rings.

The *head* is large, depressed, and blunt. It is covered by bony plates, with various ridges and furrows, and the largest of these plates is deeply divided into three narrow portions, which run along the back almost to the fin, and parallel to each other. The two *tendrils* between the nostrils, and the four under the chin, are very short; the two from the upper jaw are longer than the head, and near the root have a short membrane from the hinder edge. The *mouth* is below, and considerably behind the extremity of the nose, and is of moderate size. The *lips* are fleshy. The only *teeth* consist in a roughness on each jaw. The *eyes* are minute and high, and have circular pupils. The *gill-covers* have a scabrous skin, as have their *membranes*, which, although exposed to view, are so thick that their rays cannot be counted. There is no *slit* under the chin.

The *back* slopes to the head. The belly is prominent. The *lateral line* slopes downward, and between its commencement and the back fin are four tubercles approaching to the form of prickles. The *vent* is behind the middle.

The edge of the *back* fin is indented, and its first ray is a very strong prickle, covered with tubercles. The *pectoral* fins are far forward, under the gill-covers: each has eight rays, of which the first is a strong prickle, indented behind. Each *ventral* fin has six rays. The fin behind the *vent* is rounded. That of the *tail* ends in a crescent, the horns of which are sharp, and nearly equal in length.

23d Species.—Pimelodus conta.

A *Pimelodus* with the tail fin divided into two unequal lobes, the upper terminating in a long slender point; with eight tendrils shorter than the head; with an opaque brown scabrous body; with transparent spots on the fins; with ten rays in the fin behind the vent; and with five in the foremost of the back.

I found the *Conta* (*Khongta*) in the river Mahananda. It is an ugly animal, of little or no value, which grows to from three to five inches in length. It is long in *proportion* to its breadth, tapers gradually to the end of the tail, and is scarcely compressed. The general *colour* is a lurid brown, with some gloss of copper. The same colour extends over the fins, but these are spotted with black, and have some pellucid portions. The four larger tendrils have black rings.

The *head* is small, oval, sharpish, and covered above with bony plates variously furrowed. The hindmost plate is very deeply divided into three narrow parallel processes, which extend along the back almost to the first fin. The two *tendrils* from the nostrils, and two from the chin, are very short: the four from the corners of the mouth are longer; and each from its hinder side has a short membrane. The mouth is small, and behind and below the extremity of the nose. The *lips* are fleshy. The jaws have no perceptible *teeth.* The apertures of each *nostril* are separated only by the tendril, and are nearer the jaw than the eye. The *eyes* are far back, and high, and are very small, with circular pupils. The *gill-covers* are tuberculated, as are their *membranes,* which are so thick that the number of their rays cannot be ascertained. The *apertures* of the gills are small. There is no *slit* under the throat.

The *back* slopes very little toward the head. The *belly* is protuberant, and has two longitudinal blunt ridges, which are smooth, all the rest of the fish being rough with tubercles. The *lateral lines* are straight, and high on the sides. The *vent* is nearly in the middle.

The first ray of the foremost *back* fin is a very strong two-edged prickle, barbed before and behind. The hinder fin is minute. Each *pectoral* fin, placed under the openings of the gills, contains eight rays, of which the first is a strong prickle, indented on both edges, but most deeply behind. Each *ventral* fin has six rays, of which the first is short and undivided. The three first rays of the fin behind the *vent* are undivided, and gradually lengthen from the first, which is very short. The fin of the *tail* contains about sixteen rays, the uppermost of which is produced into a long bristle-like point.

24th Species.—PIMELODUS BOTIUS.

A *Pimelodus* with the fin of the tail divided into two equal lobes; with eight tendrils shorter than the head; with an opaque brownish scabrous body; with eleven rays in the fin behind the vent, and with six in the foremost of the back.

I have found this fish in the northern rivers of Bengal. It grows to about six inches in length, and is considered unfit for eating. It is long in *proportion* to its breadth, sharp at both ends, and scarcely compressed. Its *colour* is a lurid brown, with a little yellowness intermixed; and it is covered with a thick very rough skin. The fins are dotted.

The head is small, oval, rather blunt, and free from osseous plates, but it is covered with a very rough thick skin. All the *tendrils* are shorter than the head, and four of them proceed from under the chin, two from the nostrils, and two from the corners of the mouth. The *mouth* is below and behind the extremity of the nose. The *lips* are fleshy. The jaws are rough on the edge, but have no *teeth*. The *nostrils* are nearer the jaw than the eye. The *eyes* are high, and near each other. They are minute, and have oblong pupils. The *gill-covers* are small. Their *membranes* are so thick as to conceal the rays. There is no *slit* under the chin.

The *lateral lines* descend towards the tail. The vent is near the middle of the body. No naked bones project on the surface.

The first ray in the foremost *dorsal* fin is a prickle ending in a soft point, and smooth on the edges. The hinder fin is scabrous, and terminates abruptly behind. Each *pectoral* fin contains about eight rays, of which the first is a prickle, indented behind, and terminated by a soft point. Each *ventral* fin has five rays. The fin of the *tail* contains about eighteen distinct rays, besides small compacted ones on each side.

25th Species.—PIMELODUS NANGRA. Plate XI. Fig. 63.

A *Pimelodus* with the fin of the tail divided into two equal lobes; with eight tendrils, two of them reaching to the vent; with a smooth body, nearly opaque, and neither spotted nor striped; with nine rays in the fin behind the vent; and with eight in that of the back.

This small fish, growing to three or four inches in length, I found in the Kosi river. It is very long in *proportion* to its breadth, slightly compressed, and both above and below slopes very gradually to each extremity from the commencement of the foremost back fin. Its *colours* are rather agreeable, being silver, with some green on the back, and a faint brown streak across the foremost back fin, and another across the fin of the tail.

The *head* is sharp, long, oval, not wider than the body, and not compressed. On the nape is a small bony plate, from which two slight ridges run to the nose. The two *tendrils* from the nostrils, and the four from the chin, are nearly as long as the head, the two from the corners of the upper jaw reach nearly to the vent. The *mouth*, situated below and behind the extremity of the head, is large, and extends straight back. The *lips* are fleshy. The jaws are rough with very minute crowded *teeth*. The two apertures of each *nostril* are separated only by a tendril, and are placed nearer the jaw than the eye. The *eyes*, which are high up, are small and circular. The *gill-covers* terminate in a point. Each of their *membranes* contains about six very slender rays. There is no *slit* under the throat.

The *back* forms a ridge covered by a bony plate, which, behind, is bifid. The *shoulder-bones* are covered. The *lateral lines* are straight and high. The *vent* is behind the middle.

In the foremost fin of the *back*, which slopes much behind, the first ray is a minute bone; the second a short prickle, smooth on both edges, and ending in a substance like whalebone; the others are branched. The *pectoral* fins are low, sharp above, and shorter than the head. Each contains about nine rays, of which the first is a prickle indented behind. The *ventral* fins are smaller than the pectorals, and each has six rays, of which the first is undivided. The fin behind the *vent* slopes much backward. The first three of its rays are undivided, and gradually increase in length from the first to the third. The lobes of the *tail* fin are sharp, and contain about sixteen distinct rays, with some short compacted ones at each side.

26th Species.—Pimelodus murius.

A *Pimelodus* with the fin of the tail bifid; with eight tendrils not longer than the head; with the body opaque, smooth, and green above; with forty-two rays in the fin behind the vent; and with eight in the foremost of the back.

The *Murius (Muri vacha)* is about four inches long. I found it in the Mahananda river, and it is reckoned by the natives as belonging to the same genus with the following fish. It is long in *proportion* to its breadth, compressed at the sides, green on the back, and silver *coloured* below. The fin of the tail is blackish.

The *head* is blunt, oval, and furrowed between the eyes with a longitudinal cavity, surrounded by an elevated margin. Of the *tendrils,* none of which are longer than the head, two proceed from the nostrils, two from the corners of the mouth, and four from the chin. The *mouth* is of moderate size, and descends obliquely from the extremity of the head. There are scarcely any *lips.* The upper *jaw* is a little longer than the under. In each jaw are crowded many *teeth.* The nostrils are nearer the jaw than the *eyes,* which are large, with the longest diameter of the pupils vertical. The *gill-covers* are rounded behind. Each of their *membranes* contains about five rays.

The *back* slopes upward from the head to the foremost fin, between which organs there is a longitudinal furrow. The *belly* is prominent. The *shoulder-bones* are not visible. The *lateral lines* are straight and high. The *vent* is before the middle.

In the first *dorsal* fin the foremost ray is a prickle nearly smooth on the edges, and the last is divided to the root. The second fin is minute. Each *pectoral* fin has eleven rays, of which the first is a prickle. The *ventral* fins are much smaller, and each has six rays. The first four rays of the fin behind the *vent* are undivided, the first being very short, and the subsequent ones gradually longer. The lobes of the *tail* fin are sharp.

27th Species.—Pimelodus vacha. Plate XIX. Fig. 64.

A *Pimelodus* with the tail fin divided into two equal lobes; with eight tendrils nearly about the length of the head; with the body nearly opaque and smooth, and green above; with fifty rays in the fin behind the vent; and with seven in the foremost of the back.

The *Vacha* is common in all the larger fresh water rivers of the Gangetic provinces, grows to about a foot in length, and is an excellent fish for the table. It is long in *proportion* to its breadth, a good deal compressed, slopes gradually toward each extremity from the beginning of the first back fin, and is nearly equally prominent above and below. The *colour* above is green, with some black dots, which extend over the fin of the back; on the sides and below it is silver.

The *head* is oval, sharp, rather narrow, and smooth. Two of the *tendrils* proceed from the nostrils, two from the sides of the upper jaw, and four from the chin. The *mouth* is very large, and descends, with a little obliquity, from the extremity of the head below the eyes. There are scarcely any *lips*. The upper *jaw* is rather the longest. In both are crowded numerous small sharp *teeth*, of which there are none on the palate. The apertures of the *nostrils* are nearer the jaw than the *eyes*, which are high and of moderate size, and have the longest diameters of their pupils vertical. Each *gill-cover* consists of two bony plates, smooth, and rounded behind, and leaves exposed a thin *membrane*, containing ten rays. There is no *slit* under the throat.

The *back* is transversely rounded. The *lateral line* descends from the shoulder. The joinings of the muscles may be traced to the lateral lines, although the fish cannot be considered as diaphanous. The *vent* is before the middle.

The foremost fin of the *back* slopes very much, so that its edge is nearly vertical; and the first ray is a slender prickle, slightly indented behind, and terminated by a point like whalebone: the last is divided to the root. The hinder fin is recumbent, narrow at the root, and thickening upward. The *pectoral* fins are low, sharp above, and shorter than the head. Each contains sixteen rays, of which the first is a strong prickle, indented behind, and ending in a point like whalebone. The *ventral* fins are much smaller, and stand at a distance from each other. Each has six rays, of which the first is undivided. The fin behind the *vent* slopes to each end from the fifth ray, rapidly before, and slowly behind; the first five rays being undivided, and the others branched. The lobes of the tail fin have convex edges, and contain seventeen distinct rays, besides some compacted ones at each side.

28th Species.—Pimelodus gagata. Plate XXXIX. Fig. 65.

A *Pimelodus* with the fin of the tail divided into two equal lobes; with eight tendrils, of which two only are a little longer than the head; with a smooth opaque yellowish body; with seventeen rays in the fin behind the vent; and with eight in the foremost of the back.

The *Gagata* grows to about a foot in length, and is pretty common both in the fresh water rivers and estuaries of Bengal; but it is full of small bones, and of a very indifferent flavour. It is not very long in *proportion* to its depth, but a good deal compressed, and much more prominent above than below, and contracts somewhat suddenly at the hinder fin of the back, and at that of the vent. The back is of a purplish-brown *colour*, the sides yellow, and the belly white. The fins are edged with black.

The *head* is half oval, straight below, and a little arched above. It is blunt, and its upper part is covered by a bony plate, in most parts variously waved with intersecting lines, and extending along the back and sides in three processes. The process on each side is divided into two branches, embracing a soft oblong tumour. From the nose back to the end of this bony plate there is a cavity surrounded by a smooth ridge, and each ridge has, adjacent to its outer side, two smooth spaces. Above and below each eye there is also a ridge, the upper terminating at the nostril, and the lower at the jaw. The two *tendrils* from the nostrils, and the four from the chin, are much shorter than the head; but the two from the sides of the upper jaw are rather longer, and have a membrane extending half way along their hinder edge. The *mouth* is small, and a little behind and below the extremity of the nose. The under *jaw* has a slight ridge in the centre. The *lips* are fleshy. In both jaws are crowded many minute *teeth.* The *tongue* and *palate* are smooth. Both apertures of each *nostril* are circular, are nearer the jaw than the eye, and are separated only by the tendril.

The *eyes* are high and far back. Their longer diameters, as well as those of their pupils, are horizontal. Each *cover* of the gills consists of one smooth plate, ending behind in a point; and they do not conceal their *membranes*, which are thick. Each contains five rays. There is a *slit* under the throat.

The *back* slopes with the same curve as the head, and, before the first fin, has a bony plate, waved like that on the head. It sends a branch on each side of the fin. The *sides* slope outward toward the belly, and each above the *shoulder-blade*, which is exposed, has an oblong tumour. The *lateral line* descends from the shoulder to the end of the *tail*, which terminates in a notch. Perpendicular to each lateral line are several single rows of pores. The *vent* is behind the middle.

The fins are covered with a thick skin. The first *dorsal* slopes much behind with a concave edge, terminated by two sharp points. The first of its rays is a very short thick bone; the second is a strong prickle, smooth on the edges, and terminated by a substance like whalebone; the others are branched. The hinder fin is almost triangular, and nearly as long as that behind the vent. The *pectoral* fins are low, longer than the head, and sharp above. Each contains ten rays, of which the first is a strong prickle, barbed behind, and ending in a substance like whalebone. The *ventral* fins are much smaller than the pectorals, and each contains six rays, of which the first is undivided. The fin behind the *vent* slopes each way from the fifth ray, before very suddenly, behind more slowly; but it terminates there in a point. Its first five rays are undivided. The lobes of the *tail* fin are sharp, deeply divided, and contain eighteen distinct rays, besides compacted ones.

29th Species.—PIMELODUS MANGOIS.

A *Pimelodus* with the tail fin divided into two unequal lobes;

with eight tendrils, all a little longer than the head; with a smooth brownish olive body; with eight rays in the fin behind the vent; and with seven in the foremost fin on the back.

The *Mangois (Manggoi)* I found in the tanks of Northern Behar. It is from three to four inches in length, little compressed before, but wedge-shaped toward the end of the tail. It is every where of an uniform lurid brown *colour*, inclining to olive.

The *head* is small, blunt, fleshy, and smooth. Two *tendrils* proceed from the nostrils, four from the corners of the mouth, and two from the chin. The *mouth* extends straight from the extremity of the nose, and is large. The under *jaw* is rather the longest. The *lips* are fleshy. The only *teeth* consist of a slight roughness on the edges of the jaws. The *nostrils* are near the jaw. The *eyes* are minute, high, far forward, and near each other. The *gill-covers* are smooth, their *apertures* large, and their *membranes* exposed to view, but so thick that the number of rays cannot be readily determined; there seems, however, to be about eight in each. There is no *slit* under the throat.

There are no naked bony plates on the shoulder, nor are the *shoulder-bones* visible. There are no *lateral lines*. The *vent* is near the middle.

The first ray of the foremost *dorsal* fin is a strong prickle, smooth on the edges, and ending in a point like whalebone; the others are branched. The second fin is small and oblong. The *pectoral* fins are low, and shorter than the head, each containing eight rays, of which the first is a strong prickle, terminating in a substance like whalebone. The *ventral* fins are smaller than the pectorals, each having six rays. The first three rays of the fin behind the *vent* are undivided, and gradually lengthen from the first, which is very short. The others are branched. The lobes of the *tail* fin are sharp, the upper is the longest, and both in all contain twenty-six rays.

30th Species.—PIMELODUS GULIO. Plate XXIII. Fig. 66.

A *Pimelodus* with the fin of the tail divided into two nearly equal lobes; with eight tendrils, of which two extend to the end of the foremost dorsal fin, and four of the others are longer than the head; with an opaque smooth body, above of a dark-brown colour, with fifteen rays in the fin behind the vent; and with eight in the fore fin of the back.

The figure is taken from a bad copy of the original drawing, which was accidentally lost.

The *Gulio* (*Guli*) is found in the higher parts of the Gangetic estuaries, where the water is not very salt. It grows to about six inches in length, and is a poor fish for eating. It is long in *proportion* to its breadth, but somewhat compressed on the sides. It is more prominent above than below, and slopes to each extremity from the commencement of the first fin on the back, the slopes at the second fin, and at the fin behind the vent, being great. The *colour* above is dusky, with many dots; below it is of a dirty white; and the sides have a faint gloss of gold. All the fins are dotted.

The *head* is short, flat, and blunt. It is covered above by a smooth thinly-covered bony plate, ending behind in three short processes, and marked by an oblong cavity between the eyes. The two *tendrils* from the nostrils are shorter than the head; the two from the sides of the upper jaw extend to the end of the first fin on the back; the four from under the chin are all longer than the head; but the intermediate pair is rather shorter than the others. The *mouth* is a little below, and behind the end of the nose. The *lips* are fleshy. In each jaw are crowded numerous sharp *teeth.* There is scarcely any *tongue.* On the fore part of the palate is a bone shaped like a crescent, and covered with teeth. Both apertures of each *nostril* are nearer the jaw than the eyes: the foremost is somewhat tubular; the hinder is small and circular. The *eyes* are high, circular, and of moderate size. The *gill-covers* end behind in a blunt point, are marked with diverging rays, and leave exposed their *membranes,* which are thick, and contain many rays, the number of which, however, cannot be accurately ascertained. There is a slit under the throat.

The *back* slopes in the same slight curve with the head, and before the fin has a small thinly-covered bone, which, on each side, sends back a process. The *shoulder-bones* are thinly covered. The *lateral lines* descend from each shoulder with a curve parallel to the belly, and are porous. Several single lines of pores cross them at right angles. The *vent* is behind the middle, and has two apertures. The end of the *tail* is slightly divided into two rounded lobes.

The foremost *dorsal* fin slopes very much backward, so that its edge is almost perpendicular. The first of its rays is a very short bone, and the second a strong prickle, barbed on both sides, and ending in a substance like whalebone; the others are branched. The hinder fin is oblong, recumbent, and rounded at the end. The *pectoral* fins are low, shorter than the head, and slightly sharpened above. Each has nine rays, of which the first is a strong prickle, indented behind, and ending in a substance like whalebone; the others are branched. The *ventral* fins are smaller than the pectorals, are distant from each other; and each has one undivided and five branched rays. The edge of the fin behind the *vent* forms an arch, ending behind in a point: the four first of its rays are undivided, and gradually lengthen from the first; the others are branched.

The upper lobe of the *tail* fin is by a little the longest and sharpest. It has eighteen distinct rays, besides some compacted ones.

31st Species.—PIMELODUS MENODA.

A *Pimelodus* with the fin of the tail divided into two nearly equal lobes; with eight tendrils, of which two reach to the end of the first back fin, and four others beyond the head; with a smooth opaque body, of a dark colour above; with thirteen rays in the fin behind the vent, and with nine in that of the back.

This fish, which is found in the Kosi, Mahananda, and other rivers in the north of Behar and Bengal, has so strong a resemblance to the *Gulio* now described, that I need only mention the few particulars in which they differ, and which may perhaps be considered as insufficient to distinguish them as species, although my people from the South would not allow the *Menoda* to be the same with their *Guli*, and reckoned it a species of *Gagora*, while the *Gulio* they class with the *Tengara*.

The *Menoda* has a large black spot on the hinder fin of the back.

The head is still flatter than that of the *Gulio*.

The prickle of the back is barbed only behind.

There are five undivided, and only eight branched, rays in the fin behind the vent.

32d Species.—PIMELODUS CAVASIUS. Plate XI. Fig. 67.

A *Pimelodus* with the fin of the tail divided into two unequal sharp lobes; with eight tendrils, of which two reach beyond the tail; with a smooth nearly opaque body, brownish above; with thirteen rays in the fin behind the vent; and with eight in the foremost of the back.

The *Cavasius* (*Kavasi tenggara*) is found in the larger fresh water rivers of the Gangetic provinces, and grows to about six inches in length. It is long in *proportion* to its breadth, more prominent above than below, tapers gradually to both extremities from the commencement of the back fin, and is compressed at the sides, especially toward the end ofthe tail, which terminates in a notch. The upper parts are dotted, and are of a brown *colour*, with a green gloss, changing to purple; the lower parts are white. The form and junctions of the muscles may be discerned on the skin. All the fins, except that behind the vent, are dotted.

The *head* is oval, very little compressed, bluntish, and covered above with a bony plate, along the centre of which is a furrow. This plate is also divided into four radiated portions, the rays of the two foremost portions proceeding from their hinder edge, and those of the two hinder portions proceeding from their centres. The plate, besides, terminates behind in three narrow processes, one extending along each shoulder, and one reaching to the commencement of the back fin. The two *tendrils* from the nostrils are shorter than the head; the two from the corners of the mouth reach beyond the tail, but not to the end of its fin; of the four under the chin, the two outer reach the ends of the pectoral fins, the two inner are a little shorter. The *mouth* is small, and a little below and behind the extremity of the nose. Both apertures of the *nostrils* are nearer the jaw than the eye. The *eyes* are large and high. The *gill-covers* end in a sharp point, and conceal in part their *membranes*, each of which has about seven rays, but, on account of the thickness of the membranes, their number cannot be precisely ascertained. There is a *slit* under the throat.

4

The *back* slopes forward in the same curve with the head. The *sides* slope outward to the belly. The *lateral lines* run with a little curvature along the middle of the fish. The *shoulder-bones* are naked and sharp. The *vent*, which is near the middle, has two apertures.

The foremost fin of the *back* slopes much behind, its edge forming a concave curve. Its first ray is a short prickle, with smooth edges. The second fin is much longer than that of the vent, occupying almost the whole space between the first fin and that of the tail. Its edge forms an arch, very low before, and higher behind. The *pectoral* fins are low, shorter than the head, and sharp above. Each has about eleven rays, of which the first is a strong prickle indented behind. The *ventral* fins are rather shorter than the pectorals, and each contains six rays, of which the first are simple. From the fifth ray the fin behind the *vent* slopes both ways, before quickly, behind gently; the four first of its rays are undivided. The upper lobe of the *tail* fin is considerably the longest, has its edge more convex than that of the lower, and its end drawn out to a slender point.

33d Species.—PIMELODUS AOR. Plate XX. Fig. 68.

A *Pimelodus* with the fin of the tail divided into two very unequal lobes; with eight tendrils, of which two reach to the fin of the tail, and two to the end of the head; with a smooth body, nearly opaque, and brownish above; with twelve rays in the fin of the vent; and with nine in the foremost of the back.

The *Aor* is common in the rivers of Bengal, and in the upper parts of the Gangetic estuaries, where the water is not very salt. It grows to two or three feet in length, and is a lurid animal; but, by the natives, is considered very good eating.

It is very long in *proportion* to its breadth, and rather more prominent above than below; very slightly compressed before, but more so behind. It tapers gradually from the commencement of the back fin to both extremities; but the tail tapers somewhat suddenly at the fin behind the vent, and then enlarges toward the end, where it forms two blunt lobes. The *colour* above is purplish brown, the sides have a tinge of green, and below is white, with a silver gloss on the sides of the belly. The back fins are dotted with black, and, near the end of the hinder one, there is a fine round black spot. The lower lobe of the tail fin is stained with black and yellow.

The *head* is very blunt, long, oval, flat, and rather wider than the body. Above it is covered by a bony plate, roughened with little lines, which, behind each eye, diverge from a common centre, but, in other parts, are irregularly disposed. It has on its middle a longitudinal furrow, with a ridge from each eye to the nostril, with three short processes on each side, and a long narrow one extending up the back. The two *tendrils* from the nostrils are much shorter than the head. The two from the upper jaw extend to about the middle of the tail fin. Of the four under the chin, the two outer extend to the extremity of the gill-covers, and the two inner are a little shorter. The *mouth* is behind and below the extremity of the nose, extends straight back, and is large. The *lips* are fleshy. The *teeth*, both in the jaws and palate, are minute, sharp, and crowded. The *tongue* is smooth, and tied down. On the fore part of the *palate* is a crescent-shaped bone, covered with teeth, as are the four bones at the throat; those above are oblong; those below are parallel to each other. The foremost aperture of each *nostril* is tubular, and close to the jaw; the hindermost is circular, and nearer the jaw than the eye. The *eyes* are high and of a moderate size. The *gill-covers* consist of two plates, and are sharp behind.

They leave exposed their *membranes*, each containing fourteen rays. There is a *slit* under the throat.

The declivity of the *shoulders* is small. The *sides* slope outward. There is a small bony plate before the back fin. The *shoulder-bones* are visible. The form of the muscles is discernible through the skin. The *lateral lines* slope a little downward, and are straight. The *vent* is behind the middle.

The foremost *dorsal* fin slopes behind. Its first ray is a very short bone; the second is a prickle, smooth on the edges, and pointed by a substance like whalebone; the others are branched. The hinder fin is much longer than that behind the vent, and its edge forms an arch, rising highest behind. The *pectoral* fins are very low, shorter than the head, and sharp above. Each has eleven rays, of which the first is a strong prickle, indented behind, and ending in a substance like whalebone; the others are branched. The *ventral* fins are nearly as large as the pectorals, and each has six rays, of which the first is undivided. The fin behind the *vent* slopes quickly forward from the fourth ray, and slowly backward, and its three first rays are undivided. The lobes of the *tail* fin are sharp, and the upper is by much the longest and most pointed, and is bent in form of a sickle. In all, the fin has seventeen distinct rays, besides several compacted ones on each side.

IX. Genus.—Sisor.

Fishes of the fifth order, with a slimy skin, destitute of scales or plates on the side of the tail; with the mouth surrounded by tendrils; and with two fins on the back, the hinder one having one ray.

The *Cataphractus* of La Cepede differs in having bony plates on each side of the tail.

1st Species.—SISOR RABDOPHORUS.

Of this genus there is only one species yet known to naturalists, and this is found in the northern rivers of Bengal and Behar. It is said to grow to seven or eight feet in length, including the tail fin, and is an ugly deformed animal, on which account, perhaps, it is seldom if ever eaten. The fish is long in *proportion* to its breadth, flattened before, and tapering equally all round behind. The *colour* above is brown; with dark cloud-like marks, on the sides white, and below diaphanous.

The *head* is wider than the body, flat, blunt, and covered above with bony plates, which are marked with tubercles and ridges, and end behind in three narrow processes. There are fourteen *tendrils*, all shorter than the head: two proceed from the corners of the upper jaw; four on each side proceed from the corner of the under jaw, and are united together at their roots by a short membrane. One on each side proceeds from the vicinity of the four, which are united, and two proceed from the chin. The *mouth* is below and behind the extremity of the nose, and is small. The *lips* are fleshy. There are no *teeth*. Both apertures of each *nostril* are in the middle, between the jaw and the eye. The *eyes* are small, and very high. The *gill-covers* move freely. Each of their *membranes* has about four rays, which are seen indistinctly. Under the chin, close to the lip, is a *slit*, which resembles somewhat a second mouth.

Between the foremost dorsal fin and the head on the *back* is a bony plate, having three lobes. Near this fin, on each side, are five small plates, nearly united, and disposed in a double row. The *back*, from the end of its foremost fin to that of the tail, has a tuberculated bony ridge. The *shoulder-blades* are visible, and extend in a point above the pectoral fins; and, behind each point, are five tubercles, disposed in a row below the *lateral line*, which is straight and tuberculated. The *vent* is before the middle.

The foremost *dorsal* fin contains seven rays, of which the first is hard at the lower part, and indented behind, but soft and undivided at the point; the others are branched. The second fin has only one short prickle, with a short membrane behind. The *pectoral* fins are low, sharp above, and shorter than the head. Each has twelve rays, of which the first is a very strong prickle, indented on both edges; the others are branched. The *ventral* fins are still smaller, each containing seven rays. The fin behind the *vent* has six rays. The fin of the *tail*, ending in a crescent, has ten rays, of which the uppermost is in form of a whip or rod, and is longer than the whole body and head. The specific name has been derived from this singularity.

X. Genus.—Osmerus.

Fishes of the fifth order, with scales on part at least of the body; with no rays in the hinder of the two fins of the back, nor tendrils round the mouth; and with strong teeth in the jaws.

1st Species.—Osmerus? nehereus.

An *Osmerus?* with twelve rays in the foremost fin on the back; with fifteen in that behind the vent; and with the fin of the tail divided into three lobes.

Wana motta, Russell's Indian Fishes, Vol. II. No. 181.

This fish has a considerable affinity to the *Salmo foetens* of Linnæus and Bonnaterre, (*Tabl. Encyclop.* page 165, Pl. LXX. fig. 285,) which is the *Osmerus albidus* of La Cepede, (*Hist. des Poissons*, Tome V. page 229;) but that animal has scales all over the body, while the *Wana motta*, or *Nehare*, has them only on the hinder part of the lateral lines. Whether or not these scales are sufficient to induce naturalists to class this fish in the genus *Osmerus*, I cannot say; but the animal seems to be the link, which connects the *Osmerus* with the *Pimelodus*.

The *Nehareus (Nehare)* I found in the mouths of the Ganges. It grows to about a foot in length, and has a spine of uncommon flexibility and little elasticity. The fish is long in *proportion* to its breadth, and slightly compressed at the sides. The general *colour* is white, with many dots on the back.

The *head* is blunt and small. The *mouth*, from the end of the head, extends straight back, and is very large. The *jaws* are of nearly equal length, and do not protrude in opening. The *lips* have no bones. In both jaws are crowded many recurved *teeth*, among which some irregularly scattered are much longer than the others. Each *nostril* has two apertures, high on the head. The *eyes* are very far forward, small, circular, and flat. The *gill-covers* are membranous, with scarcely any bone, and leave large *apertures*. No rays can be distinctly traced in these *membranes*.

Both *back* and *belly* are transversely rounded. The *lateral line* extends from the eye to the end of the tail fin; and its hinder part is covered with large flexible *scales*, lying over each other like tiles. The other parts of the fish are covered with a naked smooth skin. The *vent* is behind the middle.

There are eight fins. The foremost on the *back* slopes backward, is far forward, and contains twelve rays, of which the two first are undivided, but soft, and have no membrane interposed, the first being very short; the others are branched. The hinder dorsal fin is minute. The *pectoral* fins are longer than the head, are about the middle of the side in respect to height, and are sharp above; each has twelve short rays. The *ventral* fins also are long, sharpish, and near each other. Each has nine soft rays. The edge of the fin behind the *vent* is concave, and its rays are fifteen, of which the two first are undivided, without any membrane interposed. The lobes of the *tail* fin are short, and it has about eighteen rays.

XI. Genus.—Esox.

Fishes of the fifth order, with a large mouth, devoid of tendrils, but furnished with many teeth; with one back fin, supported by rays, and opposite to that behind the vent; and with the edge of the belly blunt.

Besides the circumstances mentioned above, all the species of this genus, which I have seen in India, have the head narrower than the body, and flat above. Their eyes are high, but remote from each other, and are of moderate size, and circular. Their gill-covers are rounded, smooth-edged, and consist of one plate.

1st Species.—Esox PANCHAX. Plate III. Fig. 69.

An *Esox* with the fin of the tail entire; with the jaws equal in length, and short; with six rays in the back fin; and with fourteen in that behind the vent.

The *Panchax (Pangchak)* is very common in the ditches and ponds of Bengal, grows to about two inches in length, and lives long after being taken out of the water. It is long in *proportion* to its breadth, slightly compressed on the sides, nearly straight above, and prominent, with a regular curve below. It is of a dirty-green above, and white below; and has one silver coloured spot on the crown of the head, and another before the dorsal fin. On the lower part of this fin is a black spot, and the tail fin is edged with this colour.

The *head* is short, blunt, somewhat flattened, and half oval, being straight above and prominent below. The hinder part is covered with scales. The *mouth* is rather large, and extends straight backward from the extremity of the nose. The *jaws* are rounded, and nearly of equal length, the lower one being dotted. There are scarcely any *lips*, nor, in place of the upper, are there any bones. In each jaw are crowded many *teeth*, rather large for the size of the animal. Each *nostril* has two apertures. The *gill-covers* are scaly, and conceal their *membranes*, each of which contains four rays.

The *back* is broad, and the *belly* narrow. There is no *lateral line*. The *vent* is near the middle. For the size of the fish the *scales* are large, adhere firmly, and are rough. They are dotted on the centre, and on the edge are marked with concentric streaks.

The fin on the *back* is rounded. The *pectoral* fins are on the middle of the side, sharpest above, and as long as the head, each having sixteen rays. The *ventral* fins are smaller, and each contains six rays. The fin behind the *vent* slopes a little backward. The *tail* fin is large and oval, and has sixteen rays.

2d Species.—Esox ECTUNTIO.

An *Esox* with the fin of the tail rounded; with the lower jaw three times longer than the upper; with an angular body; with thirteen rays in the dorsal fin; and with eleven in that behind the vent.

This fish, which does not exceed a foot in length, is found in the smaller rivers and ponds of the Gangetic provinces, but is not very common. It is of a slender *make*, and tapers to both ends, is slightly compressed at the sides, and in the middle has six longitudinal angles. It is covered with scales, but is somewhat diaphanous; and on each side has a broad longitudinal stripe, shining like silver. The snout is blackish, the eyes silver coloured.

The *throat* is compressed. The *jaws* are elongated into a snout, or beak, longer than the rest of the head: the upper one is plain and sharp, and is elevated in opening the mouth; the under one is thrice as long as the under, narrow, and sharp. The *teeth* in both jaws surround the aperture of the mouth, and are rather smaller than usual in this genus. Each *nostril* has only one aperture. The *gill-covers* conceal half of the *membranes*, each of which has six rays.

The *back* is convex, and on each side is edged by a longitudinal ridge. On each side, below the lateral line, is another ridge, dividing its side into two planes. The *belly* is flat, and bounded by two ridges, which unite at the fin behind the vent. The *lateral lines* are straight, smooth, and high. The *vent* is behind the middle. The *scales* are large and smooth-edged, and fall off readily.

The fin on the *back* has its rays of nearly equal length, and some of the foremost are undivided. The *pectoral* fins are short, and sharp above; and each has nine rays, of which the first is undivided. The *ventral* fins are very small and round: each has six rays, of which the first is undivided. The fin behind the *vent* slopes backward, and the first of its rays is a short slender prickle. The fin of the *tail* contains fifteen distinct rays, besides some compacted ones at the sides.

3d Species.—Esox CANCILA. Plate XXVII. Fig. 70.

An *Esox* with the fin of the tail rounded; with both jaws elongated into a quadrangular beak; with no scales; and with seventeen rays in the fin of the back, and eighteen in that behind the vent.

The *Kuddera* A of Dr Russell (Indian Fishes, Vol. II. No. 176) resembles much the *Cancila*, (*Kangkila;*) but is distinguished by having scales on the cheeks, by different numbers of rays in its fins, and by a spot on the fin of the tail.

The *Cancila* is a very common fish in the ponds and smaller rivers of the Gangetic provinces, grows to about a foot in length, and is a very well-flavoured aliment. It is of a slender *form*, very little compressed, and tapers to both ends. It is greenish, and dotted above, and silver *coloured* or diaphanous below. On each side of the tail is a silver coloured longitudinal stripe, before and below which are four irregular black spots. The eyes are yellow.

The *head* is somewhat quadrangular, small, and covered with very little muscle, or even skin. The *mouth* is very long, both *jaws* being elongated into a straight four-sided narrow sharpish snout, with scarcely any *lips*. The under one is rather longer and wider than the upper. Along the whole length of each jaw the *teeth* are disposed in a single row, and are sharp and unequal in size. The *tongue* is tied down to the jaw. Each *nostril* has one aperture near the eye, in a triangular cavity. The *gill-covers* conceal half of their *membranes*, each containing thirteen rays.

The *lateral lines* are straight and high. The *vent* is behind the middle.

The fin of the *back* has a concave edge, and slopes backward: the two first of its rays are undivided. The *pectoral* fins are low, small, and sharp above. Each has nine rays, of which the first is undivided. The *ventral* fins are very small, each having six rays, of which the first is undivided. The fin behind the *vent* is of the same form with that on the back, and the three first of its rays are undivided. The fin of the *tail* is shaped like a fan, and has fifteen distinct rays, besides some short compacted ones at each side.

XII. Genus.—Mugil.

Fishes of the fifth order, with the body covered by scales; with two fins, supported by rays, on the back; with no detached rays before the pectoral fins; and with a longitudinal ridge in the middle of the lower jaw.

All the species of *Mugil* which I have found in the Ganges have in common the following particulars, besides those mentioned above.

The *head* is small, blunt, and in part, at least, covered with scales. The bones in the upper *lip* are minute, or altogether wanting. The *mouth* is small, or at least moderate in size, and has no *teeth*. There is no *tongue*. The *eyes* are high, and far forward, and are of moderate size. The *gill-covers* are large and scaly.

The *body*, in proportion to its thickness, is long, and covered with scales; its lower edge is rounded. The *vent* is behind the middle. There are no scale-like appendants above the pectoral or ventral fins.

All the rays of the foremost fin of the *back*, which is quite separate from the hinder, are distinctly separated by membranes, and are all undivided prickles.

1st Species.—Mugil parsia. Plate XXII. Fig. 71.

A *Mugil* with nine rays in the hinder fin of the back; with three prickles and eight soft rays in that behind the vent; with the end of the tail fin crescent-shaped; with two apertures in each nostril; with the gill-covers rounded; with a sloping head; and without stripes on the sides, which are scabrous.

This fish grows sometimes to a span in length, but is seldom above half that size. It is found in the fresh water rivers of Bengal. The length, in *proportion* to the breadth, is less than in any of the subsequent species. It is compressed at the sides, scabrous, dotted, above greenish, and beneath silver *coloured,* without stripes of different shades.

The *head* slopes gradually in the same line with the back. The *mouth* opens directly backward. The *jaws* are very nearly equal in length, the upper one having a notch to receive the ridge on the lower. Each *nostril* has two small apertures half way between the eyes and jaws. The *eyes* are a little convex, and of moderate size, and have circular pupils. Each *gill-cover* has three plates concealing the membrane, which contains six rays.

The *back* and *belly* are nearly equally prominent, and the *tail,* both above and below, contracts suddenly at the fin behind the vent. There is, properly speaking, no *lateral line,* but some radiating lines on each scale form a kind of longitudinal streaks, that somewhat represent several lateral lines on each side. The *scales,* both on the head and body, adhere firmly, and are large, dotted, and indented on the edges.

The foremost fin of the *back* is a little behind the ventrals, and slopes very much from the first ray, the edge of each interposed membrane forming a concavity. It has four slender prickles. The hinder fin is rather behind that of the vent, and slopes much backward from the second ray, which is much longer than the first, both being undivided, while the other seven are branched. The *pectoral* fins are half way up the sides, and are sharp above, and rather shorter than the head. Each contains fourteen rays. The *ventral* fins are smaller than the pectorals, each having six rays, of which the first is undivided. The fin behind the *vent* slopes backward from the third ray, which, with the first and second, are undivided.

The fin of the *tail* is rather longer above than below. It contains fourteen distinct rays, besides some compacted ones on each side.

2d Species.—Mugil cascasia.

A *Mugil* with eight rays in the second fin of the back; with two prickles and eight soft rays in that behind the vent; with the tail fin ending in a crescent; with two apertures in each nostril; with the gill-covers rounded; with a sloping head; and without stripes on the sides, which are scabrous.

This fish I found in the northern rivers of Bengal. It is usually three or four inches long, and considered as of little value for eating. It is very nearly allied to the *Parsia*, but is much longer in proportion to its height, and slopes gradually toward both head and tail from the beginning of the foremost fin on the back. It is rough to the feel, green on the upper, and silver *coloured* on the lower parts, with a faint yellowish spot on each side at the pectoral and tail fins.

The *head* is oval, and slopes forward. The *mouth* passes directly backward. The *jaws* are nearly equal in length, and smooth. The upper one protrudes in opening; the under one has a notch at the extremity. The apertures of the *nostrils* are small. The *eyes* are slightly convex, and rather small. The *gill-covers* conceal only a part of their *membranes*, each of which has about three rays.

There is no proper *lateral line*, although a longitudinal ridge on the centre of many of the scales gives an appearance of several. The *scales* are circular, adhere firmly, are of moderate size, and are indented on the edge. Towards the edge they are scabrous.

The foremost fin of the *back* has four prickles. The first ray of the hinder fin is undivided. The *ventral* fins are at a distance before the vent, and each contains one prickle and five branched rays. In the fin behind the *vent* the first of the soft rays is undivided, but jointed.

3d Species.—MUGIL ALBULA?

A *Mugil* with nine rays in the hinder fin of the back; with four prickles and nine soft rays in that of the vent; with the fin of the tail ending in a crescent; with two apertures in each nostril; with the gill-covers rounded; with a sloping head; with the sides smooth, and longitudinally striped with different shades of silver.

The number of rays in the fins of this fish does not exactly coincide with that given by authors in describing the *Mugil Albula*, (Bonnaterre, *Tabl. Encycl.* p. 180, Pl. LXXIII. Fig. 305; La Cepede, *Hist. des Poissons*, Tome V. p. 384,) and the stripes, of different shades, on the sides are abundantly evident on our fish; while it is stated, that on the *Albula* there are no stripes. The two species, however, if different, have the utmost affinity. Our fish is found in the salt water estuaries of Bengal, where it grows to a foot in length. It is very nearly allied to the *Parsia*, from the description of which it differs in only the following particulars, besides those mentioned in the specific character.

The sides are smooth, and without dots, but are longitudinally *striped* with different shades of silver, owing to the edges of the scales being paler than their centres.

There are scarcely any scales on the *gill-covers.*

On the centre of each *scale* is a slender line. The scales are angular, and very slightly indented on the edges.

The *pectoral* fins contain fifteen rays. The first ray of each *ventral* fin is a prickle. The first prickle in the fin behind the *vent* is minute.

4th Species.—Mugil cephalus?

A *Mugil* with the body longitudinally striped with black; with the eyes level; with no teeth in the jaws, the under of which is the shortest; and with the edges of the scales undivided.

Dr Russell describes a fish, which he calls *Bontah*, and considers as the *Mugil cephalus* of Linnæus. It is, no doubt, very nearly allied to what I am now describing; but the edges of its scales are indented, and there is only one prickle in the fin behind its vent.

The *Cephalus*, according to La Cepede, (*Hist. des Poissons*, Tome V. page 384,) and Bloch, (*Ichth.* Tome XI. page 129, fig. 394,) has small teeth in both jaws; so that their fish, however similar, must differ from what I am describing. According to Bonnaterre, (*Tabl. Encyclop.* p. 179, Pl. LXXIII. fig. 304,) the *Cephalus* has no teeth, and his fish may therefore be the same with that of the Ganges. None of these three writers mention whether or not the scales, in the fishes described by them, were indented on the edges. In Bloch's figure, the under jaw is represented as longest, but, in his description, the two jaws are stated to be of equal length. Considerable difficulties, therefore, still remain concerning the *Mugil cephalus*. I shall content myself with describing the fish found in the Ganges, and which the natives of Calcutta call *Sole bhanggan*, the first word being the specific, and the latter the generic term.

The *Sole* is common in the Gangetic estuaries, where the water is salt, and, near the sea, is a very rich fine-flavoured fish, from a cubit to two feet in length. It is seldom brought to table in Calcutta, because it is of a good quality only when caught near the sea, and ought to be eaten immediately after it is caught.

The fish is longer in *proportion* to its height than in the species already described, very slightly compressed, smooth, green above, and silver *coloured* below, and marked on the sides with several stripes, consisting of black dotted spots.

The *head* is level above. The *mouth* is horizontal. The *jaws* are flattened, and protrude when open, the upper one being undivided, and rather the shortest, while the under ends in a notch. There is a roughness on the *palate* and throat. The *nostrils* have double apertures, and are small, and placed in the middle between the eyes and jaw. The *eyes* are level with the head, and have circular pupils. The *gill-covers* are rounded behind, and conceal their membranes, the rays in which cannot be distinctly observed.

The *back* is convex, and slopes gradually with the head to the end of the nose. The *belly* is arched. The *tail* is suddenly contracted at the fin behind the vent, and at the hinder fin of the back. The *scales* are large, imbricated, smooth-edged, and easily displaced.

Both *dorsal* fins slope backward. The foremost, standing about the middle of the fish, has four prickles, of which the three first are strong. The hinder fin has nine rays, of which the first and second are undivided and united, the first being short; the others are branched, and the last split to the bottom. The *pectoral* fins are half way up the sides, are shorter than the head, and sharp above. Each has fifteen rays. The *ventral* fins are only a little behind the pectorals, and are shorter. Each has six rays, of which the first is undivided and short. The fin behind the *vent* is near that organ, slopes backward, and has twelve rays, of which the three first are undivided, and lengthen gradually from the first, which is short. The *tail* fin is nearly crescent-shaped, and the horns of equal length. It contains fourteen distinct rays, beside some compacted ones at each side.

5th Species.—MUGIL CORSULA. Plate IX. Fig. 97.

A *Mugil* with nine rays in the hinder fin on the back; with three prickles and nine soft rays in that behind the vent; with the tail fin ending in a crescent; with the head rising higher than the back; with the eyes protuberant; and with the sides striped with black dots.

The *Corsula (Khorsula)* is found in most rivers of the Gangetic provinces, and, in the southern parts of Bengal, has been introduced into some ponds. It grows to a foot in length, and swims with the nose and eyes above water, probably in search of insects. Although, no doubt, of a very different genus, it would seem to have much of the appearance and manners of the *Cobitis anableps*, (Bonnaterre, *Tabl. Encycl.* p. 148, Pl. LXI. Fig. 240.) It is very well tasted, and, by the Europeans, is much sought after for their tables.

The *Corsula* is very slightly compressed, scabrous; above greenish *coloured*, and below silvery. The fins are diaphanous.

The *head* rises a little from the nape to the eyes. The *mouth* is a little behind and under the nose, and extends straight backward. The upper *jaw*, which is longest, has a notch in the middle; the under is entire. The *palate* is smooth. Each *nostril* has two apertures in the middle between the eye and jaw. The *eyes* are very protuberant, and rather small. The longest diameter of their pupils is vertical. The *gill-covers* end in a point, and each contains three plates concealing the *membrane*, which has four rays.

The *back* is convex, forming a straight ridge, which slopes with a very slight declivity from the commencement of the hinder fin to the nape. The *belly* is arched. There are no *lateral lines.* The *tail* slopes suddenly at the hinder back fin. The *scales* adhere firmly, and are large, oblong, imbricated, and very minutely indented on the edges.

The fore fin of the *back* is triangular, and near the middle of the fish. It contains four prickles, rather longer than the interposed membranes. The hinder fin slopes, with a curved edge, and is much larger than the first. The two first of its rays are undivided, and the first is a short prickle: the other seven are branched. The *pectoral* fins are almost half way up the side, shorter than the head, and sharp above; and each contains fifteen rays. The *ventral* fins are a little before the foremost of the back, and are shorter than the pectorals. The fin at a considerable distance behind the *vent* is opposite to, and of the same form with, the second on the back. The first four of its rays gradually lengthen, and the first three are prickles; the nine last are branched. The fin on the *tail* contains fourteen distinct rays, besides some compacted ones at each side.

XIII. Genus.—ATHERINA.

Fishes of the fifth order, with less than eight rays in each ventral fin, and in each membrane of the gill-covers; with the palate toothless, and the lower jaw devoid of a central ridge; with the body and tail of a long form, and somewhat diaphanous; with two fins supported by rays on the back; and with a silver-coloured stripe on each side.

1st Species.—ATHERINA DANIUS.

An *Atherina* with about four rays in the foremost dorsal fin, and eight in the second; and with ten in that behind the vent.

The *Danius (Dhani)* is a minute fish, seldom exceeding an

inch in length, and is common in the Mahananda river. It has the appearance of a *Mugil*, but wants the most distinguishing character of the ridge in the lower jaw, by which that genus is marked. It is long in *proportion* to its breadth, slightly compressed, and diaphanous, with a dotted back. The belly is of a silver *colour*.

The *head* is oval, of a moderate size, blunt, level above, destitute of scales, and dotted; and on the nape has three protuberances. The *mouth* descends backward. The *jaws* protrude in opening, and have no *lips*. It has no *teeth*. The *eyes* are large, and on the sides of the head. The *gill-covers* are naked and shining. Their *membranes* contain a few rays, but so slender, that I could not ascertain the number.

The *back* and *belly* are rounded transversely. The *lateral line* is straight, and high on the side. The *vent* is near the middle. The *scales* are so thin, that they are with difficulty discernible.

The rays of the fins are so slender, that I could not exactly ascertain their number. In the fore fin of the *back* are about four, in the hinder about eight, in each *pectoral* seven, in each *ventral* six, and in the fin behind the *vent* about ten. The *ventrals* are half way between the pectorals and vent. The fin of the *tail* ends in a crescent.

XIV. Genus.—Polynemus.

Fishes of the fifth order, with some detached bristle-like rays before each pectoral fin; with scales on the head; and with two distinct fins on the back.

Besides the circumstances mentioned in the generic character, the following belong, in common, to all the fishes of this genus that I have seen in the Ganges.

Although they are not without a resemblance to the genus *Mugil*, the fishes in the genus *Polynemus* have a much greater affinity to the genera *Perca*, *Labrus*, *Sparus*, &c. of Linnæus, belonging to the fourth order, than to any included in the fifth order, which shows that this arrangement, taken from the position of the lower fins, is quite artificial.

The *body* in the genus *Polynemus* is long in proportion to its thickness, and is covered with *scales*.

The *head* is oval and small. The nose projects beyond the *mouth*, which is large, and extends directly back. There are scarcely any *lips*, but, in place of the upper, are two flat bones rounded behind. The *teeth* in both jaws are sharp, small, and crowded. Each *nostril* has two apertures. The *eyes* are covered by a thick pellucid skin or substance. The edges of the *cheek-bones* are indented. The *gill-covers* are scaly.

The *rays* before the pectoral fins are not jointed, in which this genus differs from the *Trigla*.

There are two distinct remote *back* fins. The foremost on the shoulders is supported by slender prickles, all separated by interposed membranes. The hinder fin is far back on the tail, and partly covered by scales. Each *ventral* fin contains one prickle and five divided rays. The fin behind the *vent*, and that of the *tail*, are scaly, and the latter is divided into two unequal lobes.

1st Species.—POLYNEMUS TERIA.

A *Polynemus* with four detached rays before each pectoral fin, and with the fin of the tail divided into two lobes.

Maga jellee, Russell's Indian Fishes, Vol. II. No. 183.

At one time I thought that this fish might have been the *Trigla asiatica* of Linnæus ; but the *Polynemus asiaticus* of Bonnaterre, (*Tabl. Encycl.* p. 183,) which is supposed to be the same, would appear to have only one fin on the back. For what reason I know not, La Cepede would seem to have altogether omitted this fish.

The *Teria* (*Teriya bhanggan*) is common in the Gangetic estuaries, and in the Calcutta market is often found six feet long. I have been assured, by a credible native, that he saw one, which was a load for six men, and which certainly, therefore, exceeded in weight three hundred and twenty pounds avoirdupois. It is considered by the natives as a wholesome diet, although seldom used by Europeans.

The fish, in its *form*, is compressed, and tapers to both extremities. It is scaly and rough. The *colours* are greenish above, and silver beneath, with the lower fins yellow, and the upper ones dotted. The eyes are silver coloured.

The *head* is sharp, narrow, and rough. The nose is diaphanous. The *jaws* do not protrude when open, and the under one is the shortest. The bones in lieu of the upper *lip* are scaly. On the palate are irregularly crowded numerous minute *teeth*. The *tongue* is free, blunt, very short, and smooth. On the fore part of the *palate* are three bones, covered with teeth, a small three-sided one in the centre, and an oblong one on each side. The *throat* on each side has oblong naked bones. The *nostrils* are in the middle, between the eye and jaw; the foremost of their apertures is the smallest. The *eyes* are high, and far forward. They are globular, with the longest diameters of their pupils vertical. The *gill-covers* are rounded, and smooth-edged, and each contains two plates, of which the foremost is small. Their *membranes* are half concealed, and each contains seven rays. The edges of the outer *gills* are prickly, those of the inner are covered with knobs.

The *back* forms an arch, and is transversely rounded. The *belly* is flat. The *lateral line* descends from the end of the gill-covers to the middle of the tail, and then runs straight to its end. The *vent* is before the middle. The *scales* are indented on the edge, adhere firmly, and are of moderate size. The uppermost of the rays, before each pectoral fin, is as long as the fin, the others diminish gradually in length. Above each pectoral fin is a round scale-like *appendant;* and above each ventral fin is another appendant, tapering to a point.

The foremost fin on the *back* is of a triangular form, and contains eight slender prickles, of which the first is very short. The hinder fin slopes backward, with a concave edge, and contains fifteen rays, of which the first is a short prickle, adhering closely to the second, which is undivided; the others are branched, the last being very thick, and concave behind, with scales on each margin. The *pectoral* fins are low, nearly as long as the head, and sharp above. Each contains seventeen undivided rays, of which the two first are united. The *ventral* fins, although behind the pectorals, are far forward on the fish, and are small and approximated to each other. The fin behind the *vent* is on the middle of the tail, and slopes backward with a concave edge. It has seventeen rays, of which the first and second are short prickles, closely united to the third, which is undvided; the others are branched, the last resembling the singular structure of the last ray in the hinder back fin. The upper lobe of the *tail* fin is the longest. It contains fifteen distinct rays, besides some short compacted ones at each side.

2d Species.—POLYNEMUS SELE.

A *Polynemus* with five detached rays before each pectoral fin, and extending to the vent; with no stripes on the sides; and with the lateral lines reaching to the end of the lower lobe of the tail fin.

The number of prickles in the first fin of the back, and the singular termination of the lateral line, would have induced me to suppose, that this is the *Polynemus lineatus* of La Cepede, (*Hist. des Poissons*, Tome V. page 412,) but the want of the longitudinal stripes on the sides lead me to suspect, that it may rather be his *Polynemus Emoi*, the *P. plebeius* of Gmelin, Bonnaterre, (*Tabl. Encycl.* page 183, Pl. LXXIV. Fig. 309,) and Bloch, (*Ichth.* Tome XII. page 15, Pl. CCCC.) although the last author does not represent the singular termination of the lateral line, and reckons only ten rays in the fin behind the vent.

The *Maga booshee* of Russell (Indian Fishes, Vol. II. No. 184) has certainly a strong resemblance to the *Sele* of the Ganges; but it has eight prickles in the first, and fifteen rays in the second fin of the back, and two fins behind the vent.

According to Bloch, the *Polynemus plebeius* is by the English called King-Fish, a name with which I am quite unacquainted; and it is alleged to be one of the best fishes found near Tranquebar. Our *Sele* is no doubt a light nourishing food, like most of the fishes which I have called *Bola*, but it is inferior to many of them in flavour. It is common in the estuaries of the Ganges, and is often found weighing from twenty to twenty-four pounds.

The *Sele* is of a compressed long *form*, more prominent below than above. The *colour* of the back and upper fins is bluish, of the sides and belly silver. The eyes are silver coloured.

The *head* is rather blunt, and compressed. The *jaws* do not protrude in opening; the upper is the longest. In each are crowded numerous small *teeth*. I did not examine the *palate*. The *nose* is diaphanous. The *nostrils* are in the middle, between the eye and the end of the nose, are of unequal size, and have no covering. The *eyes* are far forward, and circular. Each *gill-cover* is oval and sharp-pointed. Its *membrane* is half exposed to view, and contains five rays.

The *back* is nearly straight. The *lateral line* commences at the upper end of the gills, and runs straight to the end of the tail, where it bends downward. The *vent* is before the middle. The *scales* adhere firmly, and are flexible and indented on the edges. The longest of the *rays*, beneath each pectoral fin, reaches to the vent. Above each pectoral fin is a scale-like *appendant*, indented on the edge, and a smooth-edged one above each ventral fin.

Both the *back* fins are scaly. The foremost slopes backward, and has seven prickles. The hinder has fourteen rays, of which the two foremost are undivided and contiguous, and the others are branched, the last of them being split to the bottom. The *pectoral* fins are low, small, and sharp above; each has thirteen rays. The ends only of the *ventral* fins reach to the *vent*. The fin behind that organ slopes backward, and contains thirteen rays, of which the two first are undivided and contiguous; the others are branched, and the last of them is split to the bottom. The *tail* fin is long, and deeply divided into two unequal lance-shaped very sharp lobes, which have twenty distinct rays.

3d Species.—Polynemus risua.

A *Polynemus* with seven detached rays below each pectoral fin, the second being the longest, and being more than twice the length of the body; with seven prickles in the foremost fin of the back, and seventeen in the hinder, with fifteen rays in the fin behind the vent; and with a green back, and silver coloured belly.

The *Mango fish* of the English is one of the most delicate of those found in the Ganges, and is peculiarly excellent when in spawn, which happens in the end of spring and commencement of the rainy season, at the time when the Mango fruit is most abundant, from which circumstance the name is probably derived. I have, I think, observed three species included under this name, and Dr Russell (Indian Fishes, Vol. II. No. 185) describes a fourth; but all have exactly the same qualities and manners, nor am I sure, that the slight differences in the number of rays, which I observed, may not be accidental varieties, rather marking individual than specific difference. The variety of colour, too, may not be a specific difference, but may merely be owing to the difference of season, those that I observed in spawning season having been wholly of a bright golden colour, while those seen at other seasons were green and silver.

Dr Russell, as above quoted, considered his *Tupsee mutchee* as the same with the *Polynemus paradiseus*, nor can I take upon myself to say, that this may not be the case; but, although the accounts of the *Polynemus paradiseus*, which I have seen, are imperfect, I can venture to assert, that it is quite different from any *Mango fish* that I have observed in the Ganges.

In the figure given by Bonnaterre, (*Tabl. Encycl.* page 183, Pl. LXXIV. Fig. 308,) the *Polynemus paradiseus* is represented with filaments under the pectoral fin, which do not reach to the end of the tail fin; while in all our Mango fishes, the longest of these filaments is more than double that length. The colour of the fish, described by Bonnaterre, is golden, and the detached filament next the pectoral fin is the longest. He says, that it is found in the seas of the Indies; but, by this, he probably means the West Indies, as La Cepede and Bloch say, that the *Polynemus paradiseus* is found in the Brazils, West India Islands, and Carolina.

The number of rays in the fins given by La Cepede (*Hist. des Poissons*, Tome V. page 411) to the *Polynemus paradiseus*, differs from that found in all the *Mango fishes* which I have seen; and especially, he states the number of prickles in the fore fin of its back to be eight. Now, with us, these prickles are only seven in number; and the number of these prickles is both usually better defined, and less liable to variation, than the number of soft and divided rays. He describes the colour as green and silver.

In the figure of the *Polynemus paradiseus* given by Bloch, (*Ichth.* Tome XII. page 20, Pl. CCCCII.) the filaments reach scarcely to the vent; and he states, that the fish grows to a considerable size, which is never the case with any *Mango fish* in India. The colour, as described by this author, is green and silver.

All our *Mango fishes* are found throughout the year in the salt-water estuaries of the Ganges; but in spring, when they spawn, they ascend to the fresh water, never farther, however, so far as I know, than the tides reach. I have not seen any above nine inches in length, and their more common size is about half a foot.

The *Mango fish*, called *Risua*, (I believe from *Rishi*, a kind of holy person,) I found in the vicinity of Lukhipur. It is of a long compressed *form*, of a green *colour* on the back, and silver on the sides and belly, with a gloss changing from gold to purple. The fins are pale yellow, and those on the back are dotted with black.

The *head* is blunt, scarcely at all compressed, and covered with scales, even on the throat. The *jaws* protrude in opening. The upper is the longest, and has a notch at the end, while the under one is sharp-pointed. On each side of the *throat* are two bones covered with teeth, the upper oval, the under triangular.

The *tongue* is moveable, oval, smooth, and entire. The *nostrils* are near the eye, and their apertures are unequal. The *eyes* are high and far forward; they are small and circular. The *gill-covers* end in a point, and are flexible. They expose to view their *membranes*, each containing seven rays. The *openings* of the gills are large; and each *gill* has on the inner edge a double row of tubercles, of which the outer ones are almost cylindrical.

The *back* is arched, and transversely convex; the *sides* slope downward; the belly projects more than the back. The *lateral line* descends with an arch, and the *vent* is nearly in the middle. The *tail* is suddenly contracted at the fin behind the vent and at the hinder fin of the back. The *scales* are large, flexible, and indented on the edges, and adhere firmly. The foremost of the detached *rays*, below the pectoral fins, is about twice as long as the body; the second is a little longer, and the others shorten gradually to the seventh, which is the shortest.

The first *dorsal* fin slopes backward: the hinder fin is covered with scales, and its two foremost rays are undivided and approximated; the others are branched, and the last is split to the base. The *pectoral* fins are almost half way up the side, are longer than the head, and sharp above. Each has seventeen undivided rays, a little longer than the interposed membranes. The *ventral* fins are small. The fin behind the *vent* is in the middle between that organ and the end of the tail, and slopes backward. The three first of its rays are undivided and approximated, and the others are branched, the last being split to the base. The *tail* fin is large, and deeply divided into two sharp unequal lobes, with their points turned inward, and the upper one the longest. It contains seventeen distinct rays, besides some short compacted ones at each side.

4th Species.—Polynemus toposui.

A *Polynemus* with seven detached rays below each pectoral fin, the first being the longest, and being more than twice the length of the body; with seven prickles in the fore fin of the back, and sixteen rays in the hinder; with the same number in the fin behind the vent; and with a green back and silver belly.

This kind of *Mango fish* I found in one of the middle estuaries of the Ganges in the rainy season. The name *Toposwi*, which, in the Hindustany dialect, is given to this, as well as to the other kinds, and which Dr Russell writes *Tupsee*, is derived from the supposed resemblance of the fish to an hermit, the long rays near the pectoral fins having some appearance of the long matted hair and beard by which hermits and other pretended saints distinguish themselves from the profane. The species which I am now describing, besides the circumstances mentioned in the specific character, differs from the *Risua* in the following particulars.

The sides are faintly marked with blackish longitudinal stripes; the head and the fins on the breast are stained with red. Each pectoral fin has sixteen rays. The four first rays of the fin behind the vent are undivided.

5th Species.—Polynemus aureus.

A *Polynemus* with seven detached rays below each pectoral fin, the longest exceeding the body by more than twice its length; with seven prickles in the fore back fin, and fifteen rays in the hinder; with fourteen rays in the fin behind the vent; and with the whole colour of a golden yellow, tinged on the upper parts with dusky.

This *Mango fish* I found below Calcutta in the cold season. It differs from the *Polynemus Risua* in colour, and in the number of rays contained in the fins, as mentioned in the specific character. In both the hinder fin of the *back*, and in that behind the *vent*, the first three rays are undivided.

XV. Genus.--Mystus.

Fishes of the fifth order, with more than three rays in the membrane of the gill-covers; with the edge of the belly sharp; with the fin behind the vent very long, and united with that of the tail, and with one fin on the back.

The first species described, owing to the long rays of the pectoral fins, connects this genus in some measure with the *Polynemus*. The second species is the *Notopterus* of La Cepede, the ventral fins being very small, having escaped his notice; but, as the animal is provided with these organs, it cannot be distinguished from the genus *Mystus*.

The following circumstances, besides these mentioned in the generic character, are common to the fishes of this genus which I have observed in the Ganges.

They are shaped somewhat like the head of a lance, being much flattened at the sides, and being of a long *form*, with a considerable depth in the middle, but tapering toward each end, although more slowly toward the tail, as the greatest height is before the middle.

The *head* is small, and the under *jaw* is the shortest. On each side of the upper *lip* is a considerable flat bone. There are *teeth* in both jaws.

The *vent* is before the middle. The body in all, and part of the head in some species, is *scaly*.

1st Species.—Mystus ramcarati.

A *Mystus* with the upper jaw unarmed; with the first six rays of each pectoral fin elongated far beyond the membrane; and with ten rays in each ventral fin.

This elegant fish frequents the salt water estuaries of the Ganges, and grows to about a span in length. When living, it is everywhere of a golden colour, but soon after death this disappears. The edges of the tail fin, and of that behind the vent, are black.

The *head* is oval, compressed, devoid of scales, and sharp. The *jaws* do not protrude in opening. The bones of the upper *lip* are large, and indented on the edges. The *teeth* are minute. The *nose* extends a little way beyond the mouth. Each *nostril* has two small apertures placed close to each other, and without covering. The *eyes* are far forward. The edges of the *cheek-bones* and *gill-covers* are entire. The *membranes* of the gill-covers are short, and each contains ten rays. The *gills* reach almost to the end of the lower jaw.

The *back* slopes from the fin to the head. The ridge of the *belly* is toothed like a saw. The *lateral line* is straight. The *scales* are large, and entire on the edges. Before the fin of the back is a minute distinct prickle.

There are six fins. That of the *back* is before the middle, slopes backward, and contains fourteen rays, of which the first three are undivided, and closely connected, increasing gradually in length from the first to the third; the others are branched. The *pectoral* fins are situated low on the sides, and each contains twelve rays, of which the first six are undivided like bristles, and reach to the middle of the tail, while they are joined only at their roots by the membrane; the other six are very short. The *ventral* fins are half way between the pectorals and vent, and of a moderate size. Each has ten rays, of which the first is a prickle. The fin behind the vent is united with that of the tail, which ends in a sharp point, so as to form only one fin, which contains about one hundred and eleven rays.

2d Species.—MYSTUS KAPIRAT.

A *Mystus* with the upper jaw unarmed, and with two rays in the united ventral fins.

This fish is the *Gymnotus kapirat* of Bonnaterre, (*Tabl. Encycl.* p. 37, Pl. XXV. Fig. 83,) and the *Notopterus kapirat* of La Cepede, (*Hist. des Poissons,* Tome II. p. 190.) I have already mentioned the reason why I class it with the genus *Mystus.* I am unwilling to change the specific name adopted by two authors of reputation, although it is somewhat barbarous. Pholoe, the Bengalese name, would perhaps be more suitable.

The *Pholoe* is found in both ponds and rivers of fresh water in Bengal. In the former it is often a foot in length; in rivers it sometimes reaches to thrice that dimension. It contains too many bones to be considered good for eating. The *colour* above is green, on the sides and belly silver, with a black spot on each side, at the upper extremity of the gill-covers. The eyes are white.

The *head* is oval, rather narrower than the body, scaly, and rather blunt. On each side are four shining spaces devoid of scales. The first, above the eye, is large and oblong; the second, behind the angle of the mouth, is oblong, and indented on the edge; the third is above the second, and is kidney-formed; and the fourth, behind the eye, is small and semicircular. The *mouth* is large, and descends obliquely. The under edge of the under *jaw* is indented. The bones of the upper *lip* are quite entire. In the upper jaw are scattered some *teeth;* those in the under jaw are sharp, and disposed in a row. The *tongue* is blunt, thick, and free, with a furrow toward its point, and a margin on each edge of the root. On the tip are some large teeth disposed in rows; towards the root some small ones are scattered. Along the middle of the *palate* there is a deep furrow, with a long bone on each side; and on these bones are scattered many small teeth. Each *nostril* has only one small tubular aperture close to the jaw. The *eyes* are large, circular, and on the sides of the head. The lower edge of the *cheek-bone* is indented. The *gill-covers* are scaly, and consist of one plate, rounded behind, and disclosing the *membrane*, which contains four slender rays that cannot be distinctly seen. The bones supporting the *gills* have on their inner edge a row of spines.

The *back* transversely taken is convex, and longitudinally is arched, as is the *belly*, the edge of which is not indented. The *lateral line* commences at an oblong shining space on each shoulder, and, bending suddenly down a little way from thence, runs straight along the side above its middle. The *scales* adhere firmly, are minute, and on the upper parts of the fish are dotted.

There are only five fins. That on the *back* is near the middle, is rounded, and contains nine rays; the first very short, and, together with the second, undivided; the others are branched. The *pectoral* fins are shorter than the head, each having twelve rays. The *ventral* fins are minute, situated close to the vent, and united into one, containing two undivided rays. The fin behind the *vent*, united with that of the *tail*, is covered with scales, and has about one hundred and ten forked rays, nearly of equal lengths.

3d Species.—Mystus chitala.

A *Mystus* with the upper jaw unarmed, and with four rays in each ventral fin.

The *Chitala (Chitol)* is found in the larger fresh water

rivers of Bengal and Behar, and is usually about two feet in length, but often a half longer. The belly is uncommonly rich and well flavoured; but the back contains numerous small bones, and a strong prejudice exists against using this fish as food, owing to its being supposed to live on human carcases. In *proportion* to its length it is deeper than the other two species. Its *colour* above is green, with white transverse stripes. The sides and belly are silver, with dark dots collected into masses like clouds. There are some spots on each side of the tail.

The *head*, except on the nose, crown, and edges of the gill-covers, is covered with small scales. The bones of the upper *lip* are entire on the edges. On each jaw are crowded many *teeth*. Each *nostril* has two apertures; the foremost close to the jaw is tubular, the hinder one is small and circular. The *eyes* are far forward. The edges of the *cheek-bones* are not indented. The under plate of the *gill-covers* is slightly toothed. The *membranes* of the gill-covers are exposed to view, but are so thick that their rays cannot be distinctly numbered.

The *back* is slightly arched with a sudden declivity at the nape. The edge of the *belly* is not at all indented. The *lateral line*, commencing high, turns suddenly down at the shoulder, where is has a shining space devoid of scales, and then runs straight along the upper part of the side. The *scales* are small, and adhere firmly.

There are only five fins. That on the *back* is behind the middle, and has nine rays. The *pectoral* fins are shorter than the head, and each contains about fourteen rays. The *ventral* fins are minute, and slightly united round the vent, each having four rays. The fin behind the *vent*, united with that of the tail, contains somewhere between one hundred and twenty and one hundred and thirty rays, but the skin is so thick that the exact number cannot be ascertained.

XVI. Genus.—CLUPEA.

Fishes of the fifth order, with teeth in the jaws; with an oblong scaly body, sharp-edged beneath; with one fin on the back; and with the fin behind the vent distinct from that of the tail.

The species of *Clupea*, which I have seen in the Ganges, besides the circumstances mentioned in the generic character, have in common the following particulars:

The *body*, in proportion to its depth, is longer than in the last described genus, but is almost as much compressed at the sides, and resembles the iron of a lance. The lower edge, however, from the throat to the extremity, is very sharp, and, except where there is a fin, is toothed like a saw.

The *head* is devoid of scales, small, and much compressed. There are scarcely any *lips*; but two flat bones in some measure supply the place of the upper one.

The *body* is covered with large smooth *scales*, easily removed, and above each pectoral and ventral fin is a scale-like *appendant.*

The *pectoral* fins are low on the side. The *ventral* fins are small, and each has seven rays. The fin behind the *vent* extends almost the whole length of the tail.

1st Species.—CLUPEA PURAVA.

A *Clupea* with the hinder end of each bone of the upper lip lengthened out into a dagger-like process; with forty-six rays in the fin behind the vent; with the fin of the tail divided into two sharp lobes; and with no spots on the body.

This fish seems scarcely different from the *Peddah Poorawah*

(Great Purava) of Dr Russell, (*Indian Fishes*, Vol. II. No. 190,) although he does not notice the scale-like appendants above the pectoral and ventral fins, nor the little prickle before the fin of the back. I have derived the specific name from that given by Dr Russell, although the term *Purava* seems to be generic, as it, no doubt, forms a part of the words *Aduppa adtpuruwai*, said by La Cepede (*Hist. des Poissons*, Tome V. p. 459) to be given to the *Clupea Malabarica*. The latter fish, as described by Bloch, (*Ichth.* Tome XII. p. 116, Pl. CCCCXXXII.) is very nearly, indeed, allied to ours, although the number of rays in the fins and the colours stated by this author differ considerably from what I observed. Bloch also says, that the fish of Tranquebar does not enter rivers, while ours is one of the most common in the estuaries of the Ganges. Our fish is also nearly allied to the *Clupea setirostris* of Bonnaterre, (*Tabl. Encycl.* p. 186, Pl. LXXVI. Fig. 316.)

The *Purava* grows to about a foot in length, is so full of bones as to be little valued, and, like many fishes of this genus, dies immediately after being taken out of the water. Its back is of a green *colour*, its sides and belly silver, with a strong gloss, varying from green to purple, and with a tinge of yellow above the lateral lines. The eyes are silver.

The *head* is blunt, oval, and rather narrower than the body. The *mouth* is large, and descends obliquely from the extremity of the head. The bones, in place of the upper *lip*, extend beyond the gap into a dagger-like process, and are indented on the edges. The under *jaw* is rather the shortest, and has two ridges on its lower side. In each jaw is a single row of small sharp straight *teeth*. The fore part of the *palate* has on each side a longitudinal bone. There is scarcely any *tongue*, except a ridge, to which one end of the gills is inserted. Each *nostril* has two oblong apertures near the eye. The *eyes* are high on the head, and of moderate size, with circular pupils. The *gill-covers* are composed of three bony plates, with a sinuosity on the posterior edge of the largest and uppermost plate. Each exposes to view its *membrane*, which contains eleven rays. The bone of each *gill*, on its interior side, has a row of parallel slender teeth.

The *back* is arched, and transversely convex. The *belly* is more prominent. The *lateral line* runs straight above the middle of the side. The *vent* is behind the middle. The *scales* are orbicular. Before the back fin is a very short recurved *prickle*.

The fin on the *back*, placed between those behind and before the vent, is nearly triangular, and has thirteen rays, of which the three first are undivided and approximated; the others are branched. The *pectoral* fins are shorter than the head, and very sharp above. Each has thirteen rays, of which the first is undivided, and the others branched. The *ventral* fins are approximated, and the first ray of each is undivided. The four first rays of the fin behind the *vent* are undivided, and lengthen from the first to the fourth: the others are branched, and shorten towards the last. The under lobe of the *tail* fin is rather the longest and sharpest. In all, this fin contains eighteen distinct rays, besides compacted short ones.

2d Species.—Clupea phasa.

A *Clupea* with the first ray of each pectoral fin very long, and like a bristle; with both lobes of the tail fin sharp; with fourteen rays in the fin of the back; and with about seventy-eight in that behind the vent.

The name *Phasa* is generic; but I have appropriated it to this species, having no other to which it could be more conveniently applied. This *Phasa* is common in the brackish rivers of Bengal, and is a beautiful fish, about the size of a small herring. The snout is diaphanous, the back is of an olive-brown *colour*, the sides and belly are bright silver. All the fins are diaphanous, but that of the tail has a tinge of yellow, with a black edge on the upper lobe. The eyes are silver.

The *head* is straight, and rather sharp-edged above, and rises upward below. The end of the *nose*, which is rounded, projects considerably beyond the mouth. Both *jaws* are indented on their edges, which are sharp; nor are there any *teeth*, except those formed thus in the edges of the jaws. Each *nostril* has one opening close to the eye. The *eyes* are far forward. The edges of the *gill-covers* wave in several sinuses. The *membrane* of each contains about three rays.

The *back* slopes gradually from the head to the fin, at which it slopes downward suddenly. The *belly* is prominent. The *tail* is long, and, toward the extremity, bends down. The *lateral line* is arched slightly upward.

The first ray of the *dorsal* fin is very short, and closely joined to the second. Each *pectoral* fin has thirteen rays, of which the first is more than twice the length of the others that are branched. The first ray in each *ventral* fin is undivided. The fin behind the *vent* contains from seventy-five to seventy-eight rays, either undivided, or only once forked. The fin of the *tail* is large, and ends in a crescent.

3d Species.—Clupea telara. Plate II. Fig. 72.

A *Clupea*, with the first ray of each pectoral fin lengthened into a bristle; and with the end of the upper lobe of the tail fin blunt, as if cut across.

This fish, about the size of a small herring, is found high up in the Ganges, and in most of its larger branches and tributary streams. It is one of those which the Bengalese call *Phasa;* but I have adopted a name given to it in the Dinajpur district.

It differs little in value or appearance from the *Purava*, and, therefore, I have noted only the discriminating circumstances.

The edges of the back and tail fins are black *coloured.* In adults, the pectoral fins are black, with a green gloss.

The ends of the bones supplying the place of the upper *lip* extend but a very little beyond the corner of the mouth. The *membrane* of each gill-cover contains thirteen rays.

The *back* rises more suddenly from the head to the fin than in the *Purava.* The *vent* is before the middle. The scale-like *appendant* above the pectoral fin is divided into two. There is no *prickle* before the fin on the back.

The *back* fin is far forward, much shorter than its height, and slopes very suddenly backward. It contains fourteen rays, of which the two first are undivided, and the first is very short; the others are branched. The *pectorals* are long, the first ray, which is like a bristle, reaching to the middle of the tail. The other fourteen are much shorter, the last being undivided, and the others branched. One ray on each side of each *ventral* fin is undivided, and the intermediate ones are branched. The fin behind the *vent* contains seventy-four rays, of which the three first are undivided, and lengthen from the first, which is the shortest. The others are forked, and the last, which is the shortest, is split to the root. The fin of the *tail* is deeply divided by an obtuse sinus. The lower lobe is the longest, and is sharp; the upper is blunt, as if its end had been cut away. The distinct rays are in all nineteen, besides some short compacted ones at each side.

XVII. Genus.—Clupanodon.

Fishes of the fifth order, with no teeth in the jaws; with an oblong scaly body, sharp-edged beneath; with one short fin on the back; and with one fin behind the vent, distinct from that of the tail.

These fishes are distinguished from the *Clupeas* by no very striking circumstance, except the want of teeth, and this produces much less difference in their habits than I should have expected. Besides the circumstances mentioned in the generic character, all the species examined by me have in common the following marks:

The fish is oblong, much compressed, and *shaped* somewhat like the head of a lance, or sword-blade. The eyes are *coloured* like silver.

The *head* is devoid of scales, and more or less diaphanous before the eyes. The *mouth* is small. The *jaws* do not protrude in opening, and have scarcely any *lips;* but, in place of the upper, have two flat bones. Each *nostril*, when any is visible, has only one aperture. The *eyes* are large and high on the sides of the head. The *gill-covers* are rounded behind, consist of three or four shining plates, and expose their *membranes*, each containing five or six rays.

The edge of the *belly* is sharp throughout, and toothed like a saw. The *vent* is behind the middle.

The *back* fin is near the middle; that of the *vent* occupies a considerable proportion of the space between the vent and tail; and the number of rays in each *ventral* fin is either seven or eight. The *tail* fin has two lobes.

1st Species.—Clupanodon ilisha. Plate XIX. Fig. 73.

A *Clupanodon* with twenty rays in the back fin; with twenty-one in that behind the vent; and with eight in each ventral fin.

This is the *Sable fish* of the English in Bengal, which Dr Russell (Indian Fishes, No. 198) considered as the same with his *Palasah*, and certainly there is the utmost affinity between the fishes: but Dr Russell states the number of rays in the fins of his fish somewhat differently from what I observed; and the scale-like appendants, which the Gangetic fish has above both pectoral and ventral fins, were observed by Dr Russell only above the former. Besides, what to many will appear of more importance, the *Palasah* would seem to be very inferior in quality, as a food, to the *Ilisha* of Bengal.

Except in wanting teeth, the *Ilisha* has the most strong resemblance to the *Shad*, (*Clupea alosa*,) and there is reason to suspect that the Indian and Latin names may be radically the same. The *Ilisha* frequents the bay of Bengal and the large salt water estuaries of the Ganges, and in the rainy season ascends the larger rivers to spawn. I have seen it as high as Agra and Kanpur, but so high up it is very rare. At Patna on the Ganges, and Goyalpara on the Brahmaputra, it is pretty common, but rather poor and exhausted. About Calcutta and Dhaka it is in the utmost abundance and perfection, and is the richest and highest-flavoured fish that I know, having a taste of both the salmon and herring; but, owing to innumerable small bones, it is difficult to eat, and it is heavy of digestion. Its common size is about a foot and a half in length, but it is occasionally twice that dimension. The drawing is taken from a young fish.

The fish is above of a green *colour*, with a gloss of gold changing to purple, and below silver. In young fishes there is generally a row of four or five black spots on each shoulder; but in large individuals some of these generally, and often all of them, disappear. The fins are diaphanous; that of the tail is edged with black.

The *head* is small and oval. The *mouth* descends obliquely from the end of the nose. The upper *jaw* is rather the longest, blunt, and terminated by a deep notch. The under one is sharp, and veiled by a projecting membrane. The bones of the upper *lip* extend beyond the gap, are entire on the edges, and widen towards the end, where they are rounded. The *tongue* is free, smooth, blunt, and protuberant at the base. The *palate* is smooth. The *aperture* of each nostril is oblong, nearer the jaw than the eye, and covered by a membranous valve. The *eyes* are far forward, and covered by a thick pellucid skin. The longest diameter of the pupil is vertical. The *gill-covers* form an arch behind, are shining, and devoid of scales; and each consists of four plates. They expose to view their *membranes*, each containing six rays, of which the three uppermost are flat, the middle one terminating blunt, as if cut across. The bones of the *gills* are covered on the inside with a row of parallel teeth.

The *scales* are oblong, striated, indented on the edges, and easily rubbed off; the upper ones are dotted. The *back* and *belly* are nearly equally arched. The back before the fin forms a slight ridge. There is no *lateral line*. Above each pectoral and ventral fin is a scale-like *appendant*.

The *back* fin commences opposite to the ventrals, and extends to the commencement of the fin behind the vent. It slopes gradually backward, is devoid of scales, and the three first of its rays are undivided, lengthening gradually from the first, which is very short; the others are branched. The *pectoral* fins are much shorter than the head, low down, and sharp above: each contains fifteen rays, of which one on each side is undivided, and the intermediate ones are branched. The *ventral* fins are very small, and stand remote both from each other, and from the vent. The first ray of each is undivided, and the others are branched. The fin behind the *vent* is at a little distance both from that organ and from the end of the tail, and is covered half way up with scales, forming a furrow, in which it may be received. It slopes gradually backward, and ends in a point. The *tail* fin is covered at the root with scales, and deeply divided into two sharp equal lobes. It contains nineteen distinct rays, besides several short compacted ones on each side.

2d Species.—CLUPANODON CHACUNDA.

A *Clupanodon* with eighteen rays in the back fin, nineteen in that behind the vent, eight in each ventral fin, and five in each membrane of the gill-covers, and with the under jaw the shortest.

This fish, if it be different, has the strongest affinity to the *Kowal* of Dr Russell, (Indian Fishes, No. 186.) That fish, however, has no spot on the shoulders, and has one ray more both in the fin of the back, and in that behind the vent.

The *Chacunda* is found in the Gangetic estuaries, grows from three to eight inches long, and is little valued. It is the deepest, in proportion to its length, of our Gangetic *Clupanodons*, and its back is more prominent than its belly. Above it is of a green *colour*, with a gloss of gold, and beneath silver. The fin of the back is dotted, and that of the tail is yellowish. Behind each gill-cover, on the shoulder, is one large black spot. The eyes are silver coloured.

The *head* is of a moderate size, and sharp. The *mouth* is a little behind the end of the nose, and opens horizontally. The upper *jaw* is the longest, and has a notch at the end, and the under has a longitudinal ridge, as in the genus *Mugil*. The bones of the upper *lip* are dilated toward the end, and entire on the edges. The *nostril* is round, and near the jaw. The *eyes* are high, large, and circular, and are covered by a thick diaphanous skin: the longest diameter of the pupil is vertical. Each *gill-cover* consists of four plates, and its *mem-*

brane contains five flat rays. The bones of the *gills*, on their inner edge, have parallel slender teeth.

The *scales* are flexible, and adhere firmly. The *back* slopes in both directions from the commencement of the fin, and before this member has a sharp ridge. There is scarcely any *lateral line.* There is a scale-like *appendant* above each ventral fin.

The fin of the *back* is like a crescent, sloping backward. The two first of its rays are short prickles close to the third, which is undivided. The others are branched. The *pectoral* fins are short, and sharp above; each has fifteen rays, of which one on each side is undivided, and the others branched. The *ventral* fins are very short; one of their rays on each side is undivided, and the others are branched. The fin behind the *vent* extends almost the whole length of the tail, and is like a crescent, sloping backward. The first of its rays is a short prickle, adhering to the second, which is undivided; and the others are branched. The *tail* fin is deeply divided into two sharp lobes, and has nineteen distinct rays, besides some short compacted ones on each side.

3d Species.—CLUPANODON MANMINA.

A *Clupanodon* with fourteen rays in the fin of the back, and twenty-four in that behind the vent; and with a black spot on each shoulder.

This fish is found in most of the fresh water branches of the Ganges, and grows to about the same length with the *Chacunda*; but is reckoned better for eating. Its *form* somewhat resembles that of a sword. Its *colour* is above dirty green, and below silver, with pellucid fins, and a large black spot on each shoulder.

The *head* is of a moderate size, oval, and sharp, and a ridge bounds its sides. The *mouth*, situated behind the extremity of the nose, is small and horizontal. The upper *jaw* has a notch at the end, and is longer than the under one, which is sharp-pointed. The bones of the upper *lip* are very short, and entire on the edges. The aperture of each *nostril* is circular, and close to the jaw. The *eyes* are large and high. The *gill-covers* are rounded behind and shining, and expose to view their *membranes*, each containing five rays, of which the two uppermost are flat.

The *scales* are small, flexible, and easily removed. The *back* slopes towards each end from the commencement of the fin, before which it forms a sharp ridge. The *belly* is arched. The *lateral line* runs straight above the middle of the side, and is accompanied by a stripe of silver brighter than the rest of the side.

The *dorsal* fin slopes backward. The first of its rays is short and undivided, and the others are branched. The *pectoral* fins are short and sharp above, each containing fifteen rays. The *ventral* fins are very small, each having eight rays, of which the extreme ones are undivided. The fin behind the *vent* extends the greater length of the tail, and slopes backward. The fin of the *tail* is divided into two unequal lobes, the inferior being longest. It has nineteen distinct rays, besides some short compacted ones on each side.

4th Species.—Clupanodon chapra.

A *Clupanodon* with seventeen rays in the fin of the back, and twenty-four in that behind the vent; and with a black spot on each shoulder.

This small fish, found in the upper parts of the Ganges, so nearly resembles the species last described, that I noted merely its specific characters.

5th Species.—Clupanodon cortius.

A *Clupanodon* with fifteen rays in the fin of the back, and twenty-four in that behind the vent; and with no spot on the shoulders.

This fish, found in the Brahmaputra near Goyalpara, resembles also the *Manmina* so strongly, that I considered it unnecessary to mark any thing but the specific character.

6th Species.—Clupanodon chanpole. Plate XVIII. Fig. 74.

A *Clupanodon* withfy fteen rays on the fin of the back, and twenty-one in that behind the vent; with from three to six black spots on each shoulder; and with the ridge of the back blunt or rounded.

This fish is found in the ponds and ditches of every part of Bengal, grows to about four inches in length, and is very little valued. It is *shaped* like the wide head of a lance. Its *colours* are green above, and silver below, with from three to six black spots, extending in a line from each shoulder. The fins are pellucid, and that of the tail is dotted. The eyes are silver.

The *head* is oval, descending in the same gentle curve from the back fin to the nose, blunt, and moderate sized. The *mouth* is horizontal, opening a little below the end of the nose. The *jaws* are nearly equnl in length, and the bones, in place of the upper *lip*, do not reach beyond the gap, are club-shaped, and entire on the edges. The under *jaw* is veiled with a projecting membrane. The *tongue* is sharpish, undivided, smooth, and fixed its whole length. The *palate* is smooth, with a convexity on each side. The aperture of the *nostril* is circular, and nearer the jaw than it is to the eye. The *eyes* are globular, with circular pupils. Each *gill-cover* consists of four plates, and exposes its *membrane*, containing six rays.

The *scales* are moderate in size, smooth, and adhere firmly. On each side of the nape is a naked bony plate striated on the surface, and terminating forward in a point. The *back* and *belly* are nearly equally prominent. The *tail*, towards the end, is narrow, and bent down. There is no *lateral line.*

The *back* fin extends from above the ventral fins to the commencement of that behind the vent, and slopes backward with a concave edge. Its three first rays are undivided, and the first is very short; the others are branched. The *pectoral* fins are low, shorter than the head, and sharp above: each contains thirteen rays. The *ventral* fins are much smaller than the pectorals, each having eight rays, of which the first and last are undivided, and the others branched. The fin behind the *vent* slopes backward, and ends in a point. The two first of its rays are undivided, and the others branched. The fin of the *tail* is divided, beyond the middle, into two equal sharp lobes, and contains nineteen distinct rays, besides some short compacted ones at each side.

7th Species.—Clupanodon cagius.

A *Clupanodon* with fifteen rays in the back fin, and twenty-three in that behind the vent; with some black spots disposed in a row on each shoulder; and with the ridge of the back sharp.

This fish very much resembles the *Chanpole*, but grows to twice the size, being usually about a span in length. I found it in the rivers and ponds of Northern Behar. It is *shaped* like the head of a broad lance; and in *colour* it is above green, and below white, with an indeterminate number of black round spots in a row upon each shoulder. The eyes are silver.

The *head* is oval, moderate sized, and rather blunt. The *mouth* is horizontal. The *jaws* are nearly of equal length; the upper one has a notch at the end, and the under terminates in a point, and is veiled by a dotted membrane. The bones of the upper *lip* widen gradually downward, and are entire on the edges. The *tongue* is fixed its whole length, and is smooth and sharpish. The *nostrils* are half way between the eye and jaw. The *eyes* have oval pupils, the longest diameters of which are vertical. Each *gill-cover* contains four plates, and its *membrane* contains six rays, of which the three outer are flat.

The *scales* are middle-sized, very thin, entire on the edges, and easily removed. There are scale-like *appendants* above the lower fins; but those above the pectorals are small; and those above the ventrals are minute and indistinct.

The three first rays of the *back* fin are undivided, and the others are branched: the first is very minute, so as to be easily overlooked. The *pectoral* fins are shorter than the head, low, and sharp above: each has about fourteen rays. The *ventral* fins are still smaller than the pectorals, and do not reach to the vent: each has eight rays. The three first rays of the fin behind the *vent* are undivided, and lengthen gradually from the first; the rest are branched. The fin of the *tail* is divided into two lobes.

8th Species.—CLUPANODON MOTIUS.

A *Clupanodon* with fifteen rays in the fin of the back, and forty-three in that behind the vent; and with the edge of the upper jaw quite entire.

This fish I found in the Brahmaputra river, where it grows to be about three inches in length, and is neither in a quantity nor of a quality to be valuable. It seems to have a near affinity to the *Ditchelee* of Dr Russell, (Indian Fishes, No. 188;) but this, having the upper jaw indented on the edge, must be rather considered as a Clupea. Besides, it has only forty rays in the fin behind the vent.

The *Motius* is *shaped* somewhat like a knife, being straighter and blunter above, and more prominent and sharp-edged below. The general *colour* is silvery; but on each side is a longitudinal stripe, much brighter than the spaces above and below; and the whole is somewhat diaphanous, so that fourteen ribs on each side may be discerned through the integuments. The back and its fin, and that on the tail, are dotted. The fins are diaphanous, and that on the tail is yellow-edged, with black. The eyes are silver.

The *head* is half oval, being flat above, and rising with a curve beneath. It is small and dotted above, and on the crown has a longitudinal furrow, edged on each side with two ridges. The *mouth* is high, and descends obliquely. The *jaws* are scabrous, and the under one is the longest. The bones, in place of the upper *lip*, are as long as the jaws. The *tongue* is small, smooth, and detached. The *nostrils*, on account of the pellucidity of the head, are not to be seen. The *eyes* are convex, and have circular pupils. Each *gill-cover* consists of three flexible plates, and its *membrane* contains about five rays; but their number cannot be exactly ascertained.

The *scales* are large, very thin, and easily removed. The *back* is nearly straight, with a rounded edge. The *belly* forms an arch. The *lateral line* is scarcely, if at all, distinguishable.

The *dorsal* fin slopes backward, and its first ray is short. The *pectoral* fins are very small, and each contains fourteen or fifteen rays. The *ventral* fins are very small, and each has seven rays. The rays in the fin behind the *vent* are about forty-three, of which the three first are undivided, and lengthen gradually from the first, which is the shortest. The fin of the *tail* is divided into two lobes, of which the lower is the longest, and the distinct rays in all are eighteen, besides some short compacted ones on each side.

XVIII. Genus.—Corica.

Fishes of the fifth order, with no teeth in the jaws; with an oblong body; with the edge of the belly sharp, and toothed like a saw; with one short fin on the back; and with two behind the vent.

1st Species.—Corica soborna.

This fish is found in the Mahananda river, grows to about two inches in length, and has a very considerable resemblance to the last described species of *Clupanodon:* but the singular circumstance of two fins behind the vent prevents me from classing it in that genus. The native names *Soborno Khorika* I have adopted, the last being the generic appellation, while the specific term *Soborno* signifies golden, given on account of the little creature's beauty; for, in fact, its splendour is that of silver. It is *shaped* somewhat like a knife, being long, with the ridge of the back less prominent than that of the belly, and blunt. The greater part is diaphanous, with the bones of the skull and spine green, and the gill-covers and a longitudinal stripe on each side silver glossed with purple. The upper and lower edges, at the roots of the fins, are dotted. On the belly is an oval space, with less shining *colours.* The eyes are silver.

The *head* is oval, devoid of scales, and on the nape has a large heart-shaped protuberance, covered with dots. The *mouth* descends in a straight line from the extremity of the head. The *jaws* do not protrude in opening, are dotted, and have scarcely any *lips*; but, in place of an upper lip, are two oblong flat bones, which hang perpendicular. There are no *teeth,* nor can any *nostrils* be observed. The eyes are large, and on the sides of the head, and have circular pupils. The *gill-covers* are shining, and expose a little of their *membranes,* each of which contains about five very slender rays.

There are scarcely any *scales*, or at least they are so thin and transparent that I could not distinguish any with certainty. The *ribs* are discernible through the integuments. The *vent* is behind the middle. The *lateral line* runs straight along the upper part of the side.

The fin on the *back* is behind the middle, and contains thirteen rays. The *pectoral* fins are very small, each having twelve rays. The *ventral* fins do not reach to the vent, and each has seven rays. The first fin behind the *vent* has thirteen, and the second has four rays. The fin of the *tail* is divided into two lobes, and contains about eighteen rays.

XIX. Genus.—CYPRINODON.

Fishes of the fifth order, with an oblong, scaly, much compressed body, having both edges blunt; with one fin near the middle of the back; and with teeth in both jaws.

This genus has a very different appearance from the *Esox*, but a well defined characteristic difference is not easily discovered.

1st Species.—CYPRINODON CUNDINGA.

A *Cyprinodon* with the last ray of the dorsal fin very long.

This fish, I believe, is the *Clupea Cyprinoides* of Bloch, (*Ichth.* Tome XII. p. 24, Pl. CCCCIII.) Bonnaterre, (*Tabl. Encycl.* p. 187, Pl. LXXV. Fig. 314,) and La Cepede, (*Hist. des Poissons*, Tome V. p. 458,) although these authors, all, I suppose, copying from Broussonnet, give twenty-two rays in the membranes of the gill-covers, seventeen in the fin of the back, and twenty-five in that behind the vent.

Our fish seems undoubtedly to be the same with the *Kundinga* of Dr Russell, (Indian Fishes, No. 203.) I cannot include it in the genus *Clupea*, because the ridge of its belly is blunt; and I do not think that a good generic character can, in almost any case, be derived from the number of rays in the membranes covering the gills, not only as in many individuals this is very difficult to ascertain, but also as it differs much in fishes evidently belonging to the same genus. I, therefore, unite this fish with the *Cyprinodon* of La Cepede, as it differs just as much from the subsequent *Cyprini* as a *Clupea* does from a *Clupanodon*.

The *Cundinga* is found in the salt water estuaries of the Ganges, and is a well flavoured fish, considerably larger than a herring. It is of a long *form*, much compressed at the sides, and is dusky green on the upper parts, and silver *coloured* below.

The *head* is oval, sharpish, of a moderate size, and devoid of scales. The *mouth* is high, descends obliquely in a straight line, and is large. The *jaws* have scarcely any lips, and the under one is considerably the longest. In place of an upper *lip* are two very long, blunt, flat bones, entire on the edges, descending before almost vertically, and then bending back to a horizontal position. In both jaws are crowded many minute *teeth;* and the same is the case on the upper surface of the *tongue*, which is blunt, undivided, thick, and detached below. On the fore part of the *palate* are three bones covered with teeth. The intermediate one, standing transversely, is oblong; the side ones, standing longitudinally, widen at one end, and narrow at the other. Each *nostril* has two oblong apertures, mid-way between the eyes and jaw. The *eyes* are very large, round, and convex, with oval pupils, the small ends pointing downward. Each *gill-cover* is rounded behind, and consists of three plates. It exposes one half of its *membrane*, containing twenty-five rays. The bones supporting the *gills* on their inner edges have a row of bristles.

The *body* is covered with large dotted scales, adhering firmly. The *back* is less prominent than the *belly*, which slopes rather suddenly at the fin behind the vent, and is blunt, and quite entire on the edge before the vent. The *lateral line* is above the middle, and is rather curved, with the concavity upward. The vent is behind the middle. Above each pectoral and ventral fin is a scale-like *appendant*.

The *back* fin is rather behind the middle, and slopes backward, having nineteen rays, of which the first five are undivided, and lengthen gradually from the first, which is very short: the next thirteen are branched, and the last is twice as long as any of the others, undivided, jointed, and flat, being shaped like the head of a narrow lance. The *pectoral* fins are low, and are shorter than the head: each contains fifteen rays, of which the first is undivided, and the others branched. The *ventral* fins are short, and approximated, and each contains one undivided and nine branched rays. The fin behind the *vent* is crescent-shaped, and contains twenty-six rays, of which the six first are undivided, and lengthen gradually from the foremost, which is very short; the others are branched, and the last split to the root. The fin of the *tail* is deeply divided into two sharp lobes, of which the upper is the longest. It contains nineteen distinct rays, besides some small compacted ones at each side.

N. B.—Perhaps the third species of the following genus should rather have been placed here?

XX. Genus.—Cyprinus.

Fishes of the fifth order, with one back fin, which has rays, but both this and the pectoral fins are unarmed; with no teeth in either jaw; with the edge of the visceral cavity blunt, in part at least; and with four tendrils, at the utmost, near the mouth, but this organ is often altogether destitute of these appendages.

This very numerous class of fishes has been usually subdivided from the form of the tail fin, and from the absence or presence of fewer or more numerous tendrils; but these organs, in several species, are so minute as to render it uncertain whether some individuals of the same species have them or not; and the fin of the tail, in all the Gangetic species except one, is divided into two lobes, more or less distinct. Besides, this arrangement brings together species that have little general resemblance. I have, therefore, been under the necessity of adopting other divisions, which, indeed, bring together the species most nearly allied; but, I must confess, that some of the divisions are not distinguished by very strictly defined characters. In order to save repetitions in the specific characters, I have given each division a common name, derived generally from some native generic appellation.

Besides the circumstances mentioned in the generic character, the species of *Cyprinus* found in the Ganges, except in a few cases that will be specified, have in common the following marks:

The *head* is devoid of scales, and has shining bony *gill-covers*, which move freely, and leave a large *opening* on each side for the gills. Each *nostril* has two apertures separated merely by a thin membrane, which projects from the surface, and serves as a valve for one or other. In a few of the smaller species, I have not been able to trace this structure; but, in the larger species, it may be considered as existing, when the contrary is not mentioned in the description. The *membrane* of the gill-cover very generally contains three rays, and never more than four. The former is to be supposed as the case wherever the contrary is not mentioned.

I. Division.—CYPRINUS CHELA.

Fishes of the Genus *Cyprinus*, with a sharp edge on the space between the ventral fins and the fin behind the vent.

The *Cyprini* of this division differ from the *Clupanodons* in the smaller number of rays in the membranes of the gill-covers, and in the fore part of the belly having a blunt edge; but on the whole, the *Clupanodons*, and these *Cyprini*, are but indifferently distinguished. To this division belong the *Cyprinus cultratus* and *C. clupoides* of Ichthyologists. Besides the characters belonging to the genus and division, the seven following species have in common the following marks:

The *form* is long, much compressed, and somewhat like the head of a lance, or blade of a knife. The eyes are silver *coloured.*

The *head* is small, and devoid of *tendrils.* The *eyes* are high and far forward. The *back* fin is considerably behind the middle. The *ventral* fins are at some distance before the vent.

1st Species.—CYPRINUS CACHIUS.

A *Cyprinus* of the *Chela* kind, with three rays in each ventral fin; with about twenty-six in that behind the vent; and with about seven in that of the back.

The *Cachius (Kachhi)* is a minute fish, scarcely an inch long, and is found in the Ganges, about the commencement of the Delta. It is of an oblong *shape,* more prominent below than above. The greater part is diaphanous, with a gloss changing from green to purple. The back and belly are dotted, the upper part of the head is green, on each side is a silver stripe, and the membrane investing the visceral cavity is of the same *colour.* The fins are diaphanous, that of the back being dotted.

The *head* is blunt, and half oval, being flat above. The *mouth* descends from above, and is of a moderate size. The *jaws* do not protrude in opening, and the under is rather the longest. The upper *lip* has no bones. The *nostrils* are half way between the eye and jaw. The *eyes* are large, with circular pupils. I am not certain as to the number of rays in the membranes covering the gills, but think that I can count three in each.

Over the silvery membrane investing the visceral cavity, the *ribs* are visible. From the commencement of the fin, the *back* slopes both ways very gently, and its edge is blunt. The *belly* is more prominent, the *sides* are perpendicular, and the *vent* is before the middle. The *lateral line* runs straight above the middle of the side. The *scales* are of moderate size, but are so transparent as to be scarcely visible.

The *pectoral* fins are low, and longer than the head, and each contains about eight rays. The first ray of each *ventral* fin is a very long bristle, and the other two are minute. The fin of the *tail* is divided into two lobes.

2d Species.—CYPRINUS ATPAR.

A *Cyprinus* of the *Chela* kind, with five rays in each ventral fin; with twenty-five in that behind the vent; and with nine in that of the back.

The *Atpar* I have found in the two most remote branches of the Ganges, the Brahmaputra and Yamuna, (Jumna R.) and I have also found it in ponds near the former. It grows to about three inches in length, and is nearly diaphanous, with a gloss on the sides, changing from green to purple. The head, a longitudinal stripe on each side, and the membrane lining the visceral cavity, are silver *coloured;* and over the latter, nine ribs on each side may be distinguished through the integuments.

The fins are yellowish, and that of the back, together with the upper ridge of the fish, are dotted.

The *head* is oval. The *mouth* is of moderate size, and descends almost perpendicularly. The *jaws* do not protrude in opening; and the under one is rather the longest. There are no bones in the upper *lip*. The *nostrils* are nearer the eyes than the jaw. The *eyes* are of moderate size, and convex, with circular pupils. The *gill-covers* are rounded behind, and each consists of only one plate, exposing its *membrane* to view.

The ridge of the *back* is blunt, and slopes a little both ways from the commencement of the fin. The *lower edge* is more prominent. The *vent* is nearly in the middle. The *lateral line* is low, and in a curve parallel to the belly. The *scales* are small, and adhere firmly.

The first ray of the *dorsal* fin is undivided, the remainder branched, and the last of them split to the root. The *pectoral* fins are much longer than the head, and each has eleven or twelve rays. The first ray of each *ventral* fin is a very long bristle, and the others are minute. The fin behind the *vent* occupies most of the under side of the tail; and the first three of its rays, increasing from the first, which is short, are undivided; the others are branched, the last of them being split to the root. The *tail* fin, with a deep sinus, is shaped like a crescent, having both horns sharp, and nearly equal in length.

3d Species.—Cyprinus laubuca.

A *Cyprinus* of the *Chela* kind, with seven rays in each ventral fin; with twenty-four in that behind the vent; and with ten in that on the back.

The *Laubuca* (*Layubuka*) I found in the ponds of the northern parts of Bengal. It is from two to three inches long, most prominent below, and of a brownish green *colour* above, and silver beneath, with a gloss on the sides changing from green to purple. Above the insertion of each pectoral fin is a greenish spot, and at the end of the tail there is another. The fins are diaphanous and pale green, and, with most parts of the fish, are covered with black dots.

The *head* is sharpish, half oval, sloping less above than below, and rather wider than the body. The *mouth* is small, and descends obliquely. The *jaws* do not protrude in opening, and are nearly equal in length; but the under one is by a trifle the longest. They are devoid of *lips*, but, in place of the upper, are two minute bones entire on the edges. There is a slight roughness on each jaw, which perhaps might give the fish a place in the genus *Cyprinodon*. Each NOSTRIL has one aperture in the middle between the eye and jaw. The *eyes* are large and globular. The *gill-covers* consist each of three plates, end in a point, and conceal their *membranes*.

The *back* slopes gently each way from the commencement of the fin. The *sides* are convex. The *belly* forms an arch. The *lateral line* descends obliquely with a curve parallel to and near the belly. The *scales* are large, and easily removed.

The fin on the *back* slopes very little. Its first ray is short, and, together with the second, is undivided; the others are branched; and the last is split to the root. The *pectoral fins* are on the middle of the side, and are longer than the head, each having twelve rays. The *ventral* fins are distant from each other, and the first ray of each is undivided, and the others branched. The fin behind the *vent* is more distant from that organ than from the end of the tail, and its edge forms an arch. Its first ray is very short, and, together with the two following, is undivided; the others are branched. The fin of the *tail* is large, and divided into two lobes.

4th Species.—Cyprinus phulo.

A *Cyprinus* of the *Chela* kind, with eight rays on each ventral fin; with twenty in that behind the vent; and with nine in that of the back.

The *Phulo* (*Phul chela*) is found in the rivers and ponds of the north-eastern parts of Bengal, grows to about three inches long, and is little valued. It is more prominent on the lower edge than above, where it is green, with silver *coloured* sides. The fins of the tail and behind the vent are tinged with yellow.

The *head* is oval, and a little sharpish. The *mouth* is small, and opens horizontally. The *jaws* protrude in opening, and have no *lips*, but, in place of the upper, are two very minute bones. The upper jaw is rather the longest; the under one is blunt. Each *nostril* has only one circular opening near the eye. The *eyes* are circular. The *gill-covers* do not conceal their *membranes*.

The *back* is nearly straight, with a blunt edge. The *belly* is prominent and arched. On each side are two *lateral lines;* one high up, and straight, the other bent downward parallel to the edge of the belly. Above each pectoral and ventral fin is a scale-like *appendant*. The *scales* are minute.

The *back* fin slopes behind; the first of its rays is very short, and, together with the second, is undivided; the others are branched, the last being considerably longer than that next to it. The *pectoral* fins are low, and longer than the head, and each contains eleven rays. The *ventral* fins are very short. The first ray of the fin behind the *vent* is short, and, with the two next, is undivided, the others being branched. The fin of the *tail* is divided into two lobes, of which the under is the longest; and it has eighteen distinct rays, besides some short compacted ones on each side.

5th Species.—Cyprinus gora.

A *Cyprinus* of the *Chela* kind, with eight rays in each ventral fin; with seventeen in that behind the vent; with nine in that of the back; with the lateral line devoid of spots; and with some teeth like those of a saw on the edge of the belly.

The *Gora (Ghora Chela)* is one of the largest fishes in this division of the genus, and is commonly about a span in length. I found it in the Brahmaputra, near Goyalpara.

The *head* is bluntish, and on the upper part has scales as far as the nostrils, but is naked on the sides. It is half oval, being flat above and sloping below, and is narrower than the body. The *mouth* is large, and descends obliquely. The upper *jaw* is protruded in opening, and, in place of a *lip*, has two small bones entire on the edges. The under jaw has a sharp point turned upward. The *nostrils* are very near the eye. The *gill-covers* expose their *membranes.*

The *back* is straightish, and has a blunt edge. The *belly* is more prominent. On each side are two *lateral lines*, the upper straight, the under parallel to the edge of the belly, with a break in its middle. Above each pectoral and ventral fin is a scale-like *appendant.* The *scales* are small, thin, and well fixed.

The *dorsal* fin slopes behind; the first of its rays is short, and, with the second, is undivided; the others are branched, and the last of them split to the root. The *pectoral* fins are very low, and longer than the head. Each has about thirteen rays. The *ventral* fins are very small. The first three rays of the fin behind the *vent* are undivided, and closely united; and the others are branched, the last being divided to the root. The *tail* fin is divided into two lobes.

6th Species.—Cyprinus morar. Plate XXXI. Fig. 75.

A *Cyprinus* of the *Chela* kind, with eight rays in each ventral fin; with twelve in that behind the vent; with ten in that of the back; and with the edge of the belly smooth.

The *Morar* (*Morur*) I have found in the rivers Yamuna and Tista, two large but distant branches of the Ganges. It is about a span in length, higher flavoured than any species in this division, somewhat resembling a smelt, and, by the natives, is sought after as a delicacy. Its *colour* is silver, with a green back, and all the fins clear, except that of the tail, which is yellowish, and stained with black.

The *head* is oval, and blunt. The *mouth* is very small, low, and horizontal. The *jaws* protrude in opening; and the upper is the longest, has a fleshy *lip*, and is undivided at the end. Each *nostril* has only one aperture, near the eye, which is of moderate size, and has the longest axis of the pupil in a vertical position. The *membranes* of the gill-covers are exposed to view.

The edge of the *back* is blunt. Each side has two *lateral lines;* the upper straight, the under parallel to the edge of the belly. Above each pectoral and ventral fin is a scale-like *appendant.* The *scales* are large, and readily separate.

The three first rays of the *back* fin are undivided and smooth, the first of them being minute; the others are branched, and the last split to the root. The *pectoral* fins are low down, and shorter than the head; each has fifteen rays. The *ventral* fins are small. The fin behind the *vent* commences near that organ, and the three first of its rays are undivided, and the others branched, the last being split to the bottom. The lobes of the *tail* fin are well marked, and contain about twenty distinct rays, besides some that are short and compacted.

7th Species.—Cyprinus bacaila. Plate VIII. Fig. 76.

A *Cyprinus* of the *Chela* kind, with nine rays in the ventral fins; with sixteen in that behind the vent; with nine in that of the back; and with all the fins clear as water.

The *Bacaila* of Hindustan Proper is one of the most generally diffused fishes in the fresh water rivers of India, as I have found it not only in all the Gangetic provinces, but in the heart of the Peninsula. It grows to about six inches in length, but is little valued, being very insipid. The *colour* is above dusky green, below silver. Sometimes, but not always, there is a large cloud-like spot on each shoulder; and there always are many small well-defined spots along the upper lateral lines.

The *head* is bluntish, narrow, and half oval, being flat above, and curved upward below. The *mouth* is large, and descends obliquely. The *jaws* protrude in opening, and have no *lips;* but, in place of the upper, there are two small bones about the length of the jaw. The upper jaw ends in a notch; the under, which is the longest, terminates in a turned-up point. The *tongue* is sharpish, entire, smooth, and tied down its whole length. The *palate* is smooth, and on each side is protuberant. Each *nostril* has only one circular aperture, very near the *eye,* which is globular. Each *gill-cover* has three plates, is rounded behind, and leaves exposed its *membrane.* The bones of the *gills,* on their inner edge, have a row of parallel bristles.

The *back* is nearly straight, with a rounded edge. The *belly* projects with a more considerable curve. Each side has two *lateral lines,* the upper straight, the under parallel to the edge of the belly. Above each ventral and pectoral fin is a scale-like *appendant.* The *scales* are small, thin, and adhere firmly.

The *back* fin slopes behind: the first of its rays is short, and, with the second, is undivided; the others are branched; and the last of them split to the bottom. The *pectoral* fins are very low, and are rather longer than the head: each has twelve rays, of which the first is undivided, and the others branched. The *ventral* fins stand apart from each other, and are shorter than the pectorals; and both the extreme rays of each are undivided, and the intermediate ones branched. The fin behind the *vent* slopes backward; and the two first of its rays are undivided, and the others branched. The fin of the *tail* has its upper lobe the shortest, and contains nineteen distinct rays, besides some that are short and compacted.

II. Division.—Cyprinus barilius.

Fishes of the Genus *Cyprinus*, with the body very long and much compressed; with the sides irregularly marked with numerous incomplete transverse bars or spots; with the back fin placed considerably behind the middle of the fish; and with one lateral line at least running parallel to the lower edge of the fish.

The fishes of this division, on account of the numerous spots or marks on their sides, have a considerable resemblance to our Trout, and, by the English in Bengal, some of them have been considered as belonging to that genus; but they have only one fin on the back. I have not been able to trace in authors any species that should be referred to this division;—all the *Cyprini* described as having spots being much wider in form, and having the lateral lines straight, and the back fin nearer the middle of their body.

Several species of *Cyprinus* included in the first division, and nearly allied to the *Clupanodons*, approach very near those resembling trouts, as they have a lateral line parallel to their lower edge, and their back fin far behind their middle; but the marks on the sides of the second division form a well-defined character. Besides the above circumstances, and those mentioned in the generic character, and in that of the division, I observe, that all the species of this division in the Ganges have in common the following circumstances :

All the marks on their sides, whether bars or spots, are of a black *colour.* The upper parts are green ; the under parts and eyes are silver.

The *head* is oval. The *eyes* are high, on the sides of the head, and have circular pupils.

Both edges of the *body* are blunt. The *vent* is behind the middle.

The *dorsal* fin slopes backward : its two first rays are undivided, and closely united, the first being the shortest ; the others are branched, and the last divided to the root. The *pectoral* fins are shorter than the head, and low. The *ventral* fins are still smaller, and each contains nine rays, except in one species, where there are only eight. The fin behind the *vent* slopes backward, and its last ray is divided to the bottom. The *tail* fin has two lobes, except in one species, where it is like a crescent.

8th Species.—CYPRINUS BARILA.

A *Cyprinus* of the *Barilius* kind, devoid of tendrils, with incomplete bars on each side ; with eight rays in the fin of the back ; and with thirteen in that behind the vent.

The *Barila* is common in the rivers of Northern Bengal, grows to about three inches in length, and is little valued. The *form* somewhat resembles the blade of a knife. Its *colour* has a yellow tinge on the fin of the tail, and on that behind the vent. The bars are disposed in a row on each side.

The *head* is sharp and small. The *mouth* is small, and descends obliquely in a straight line. The *jaws* have scarcely any lips, and protrude in opening : the upper is the longest, and is entire at the end. The *nostrils* are near the *eyes*, which are large and flat. The *membranes* of the gill-covers are not concealed.

The *scales* are large, easily removed, and entirely conceal the shoulder-blades. The *back* is nearly straight, and the *belly* is more prominent. There are two *lateral lines,* the upper of which is straight.

Each *pectoral* fin has fourteen rays, and each *ventral* only eight. The first ray in the fin behind the *vent* is short and undivided. The under lobe of the *tail* fin is the longest.

9th Species.—Cyprinus chedrio.

A *Cyprinus* of the *Barilius* kind, devoid of tendrils, with incomplete bars on each side; with nine rays in the back fin; and with thirteen in that behind the vent.

The *Chedrio (Chedri)* is found along with the Barila, from which it only differs in the number of rays in the fin of the back. I should have considered this as a mere accidental variety, but the fishermen held them to be distinct. The first ray of the *back* fin is very short, and both it and the next are undivided.

10th Species.—Cyprinus barna.

A *Cyprinus* of the *Barilius* kind, devoid of tendrils, with incomplete bars on the sides; with nine rays in the back fin, and twelve in that behind the vent; and with a golden stripe along the upper lateral line.

The *Barna* I found in the Yamuna and Brahmaputra rivers, the extreme branches of the Ganges. It grows to about three inches in length, and is frequently observed swimming close by the shore on sloping sandy banks, on which account it is, in some places, called *Bali Bhola,* or the *Sand Bhola.* It has such a strong affinity to the two last mentioned species, that, in many parts, it is included with them in the same generic name *Bareli;* and is chiefly distinguished by having smaller scales, and the golden line on each side. It has much the *shape* of a knife-blade. The three hinder fins are of a yellow *colour*, and those of the back and tail are stained with black, especially on the edge.

The *head* is small and sharpish. The *mouth* is large, and descends obliquely. The upper *jaw* protrudes in opening. The *nostrils* are in the middle between the eyes and jaw. The *eyes* are large. The *membranes* of the gill-covers are exposed to view.

The *scales* are of a moderate size. The *back* is nearly straight, and the *belly* is more prominent. On each side there are two *lateral lines*, the upper of which is straight.

Each *pectoral* fin has about thirteen rays. The *ventral* fins do not reach the vent. The first ray of the fin behind the *vent* is short, and closely joined to the second, which, like it, is undivided: the others are branched. The fin of the *tail* contains nineteen distinct rays, besides short compacted ones.

11th Species.—Cyprinus vagra.

A *Cyprinus* of the *Barilius* kind, devoid of tendrils, with incomplete bars on the sides; with nine rays in the fin of the back; and with fourteen in that behind the vent.

The *Vagra* I found in the Ganges about Patna, where it grows to about three inches in length. It is *shaped* somewhat like the blade of a knife, with transverse bars descending to the lateral line. The fin of the tail is yellow, edged with black.

The *head* is somewhat blunt and small, as is the *mouth*, which descends a very little in a straight line. The *jaws* are protruded in opening, and nearly equal in length, the upper being undivided at the end. The *nostrils* are in the middle between the eye and jaw. The *eyes* are far forward, with circular pupils. The *gill-covers* are rounded behind.

The *scales* are small. The *belly* is more prominent than the *back*. Above each pectoral and ventral fin is a large scale-like *appendant*.

Each *pectoral* fin has thirteen rays. The first three rays of the fin behind the *vent* are undivided. The lobes of the fin of the *tail* are equal, and have sixteen distinct rays, besides several short compacted ones.

12th Species.—Cyprinus bendilisis. See my Journey in Mysore, Vol. III. Pl. XXXII.

A *Cyprinus* of the *Barilius* kind, with two tendrils; with incomplete bars on the sides; with nine rays in the back fin; and with eleven in that behind the vent.

This fish I found in the rivers of Mysore, where it grows four or five inches long. It is *shaped* somewhat like the head of a lance; but the edge of the belly projects a little more than that of the back. The bars descend almost to the lateral line. The scales of the back toward the root are black. The fin of the tail is yellow, edged with black.

The *head* is sharp and small. The *mouth* is small, and descends obliquely. The *jaws* protrude in opening, the upper being the longest and widest. At each angle of the mouth is a very short *tendril*. Each *nostril* has only one aperture near the *eye*, which is flat. The *gill-covers* are rounded behind, and each contains three plates.

The *scales* are large and firmly fixed. The *belly* slopes suddenly at the fin behind the vent. The *shoulder-bones* are bare, and shine bright. There are no scale-like *appendants* above the ventral fins.

Each *pectoral* fin has thirteen rays. That behind the *vent* commences close to this organ; the three first of its rays are undivided, and lengthen gradually from the first, which is very short; the others are branched. The under lobe of the *tail* fin is the longest, and it has in all nineteen distinct rays, besides some short compacted ones.

13th Species.—Cyprinus shacra.

A *Cyprinus* of the *Barilius* kind, with four tendrils; with incomplete bars on the sides; with nine rays in the fin of the back; and with ten in that behind the vent.

This fish differs very little from the last, except in having four tendrils in place of two. I found it in the river Kosi, where it grows to four or five inches in length. It is *shaped* like the head of a lance. Numerous transverse bars descend to the lateral lines. The fin of the tail is yellow, which, on the lower lobe, is stained with black.

The *head* is bluntish and small. The *mouth* is of moderate size, and horizontal. The *jaws* protrude in opening, the upper being the longest, and being undivided at the end. On each side a *tendril*, shorter than the head, proceeds from the middle of the upper jaw, and another from the corner of the mouth. Each *nostril* has only one aperture in the middle, between the eye and jaw. The *eyes* are far forward. The *gill-covers* are rounded behind.

The *scales* are moderate sized, and easily separated. The *shoulder-blades* are naked and shining. The pectoral and ventral fins have no scale-like *appendants*.

Each *pectoral* fin has thirteen rays. The two inner rays of each *ventral* fin are almost united. The first ray of the fin behind the *vent* is short, and both it and the second are undivided; the others are branched, and the last is divided to the root. The under lobe of the *tail* fin is the longest, and it contains twenty distinct rays, besides some short compacted ones.

14th Species.—Cyprinus cocsa. Plate III. Fig. 77.

A *Cyprinus* of the *Barilius* kind, with four tendrils; with both incomplete bars on the sides, and a spot on the middle of each scale; with nine rays on the fin of the back; and with eleven in that behind the vent.

The *Cocsa* (*Khoksa*) is found in the northern rivers of Bengal and Behar, especially in the Mahanunda. It grows to about a span in length, and is *shaped* like the head of a lance, but is rather more prominent below than above. The short bars on its sides disappear in the dead fish. There is a row of small oblong spots on each side of the lateral line. The fins are yellow, but, in the dead fish, become white, which is rather uncommon, the contrary change of *colour* being much more usual. The lower lobe of the tail fin is stained with black.

The *head* is small and sharpish. The *mouth* is small, and descends obliquely from the extremity of the head. The *jaws* protrude a very little in opening, and the under one is rather the shorter, and is smooth. On each side are two *tendrils* much shorter than the head, one from the corner of the mouth, the other from the side of the upper jaw. There are scarcely any *lips*. The *nostrils* are near the *eyes*, which are of moderate size. The *gill-covers* are rounded behind, and edged with a pellucid membrane. Their proper *membranes* are exposed to view.

The *scales* are large, and dotted near the edge. The *shoulder-blades* are naked and shining. There is a small scale-like *appendant* above each ventral fin.

Each *pectoral* fin contains about thirteen rays. The *ventral* fins do not reach to the *vent*, near to which, however, the fin behind it comes. This slopes with a concave edge, and its two first rays are undivided, the first being very short, the others are branched. The lobes of the tail fin are sharp, and contain nineteen distinct rays, besides some short compacted ones.

15th Species.—Cyprinus chedra.

A *Cyprinus* of the *Barilius* kind, devoid of tendrils, with a spot on the bottom of each scale upon the sides; with ten rays in the back fin; and with eleven in that behind the vent.

The *Chedra* is found in the northern rivers of Bengal, grows to about six inches in length, and is little valued. It is deeper in the *form* than any other in this division. In the position of the spots it agrees with the last described species, and with that immediately following, and differs from the subsequent fishes, each spot in these being diffused over a portion of more than one scale. The dorsal and pectoral fins are dotted, the former brownish, the latter white like the ventrals. The fin behind the vent and that of the tail are reddish, the latter inclined to brown.

The *head* is of moderate size and blunt. On the nose are many blunt tubercles. The *mouth* is large and straight, and commences at the extremity of the head. The upper *jaw* does not protrude in opening; the under one on the outside is rough, with sharp thick-set tubercles. There are scarcely any *lips*. The *eyes* are small. The *gill-covers* are rounded, and do not conceal their *membranes*.

The *scales* are large, adhere firmly, and are rough on the surface with little blunt grains. The *shoulder-bones* are naked, large, shining, and dotted. Above each ventral fin is a scale-like *appendant*.

The *pectoral* fins are rounded at the end, and each contains about fourteen rays, of which the first is broad, but soft and undivided. The *ventral* fins scarcely reach to the *vent*. The three first rays of the fin behind this organ are undivided, and lengthen from the first, which is short: the others are branched. The *tail* fin ends like a crescent, and contains about eighteen distinct rays, besides several small compacted ones.

16th Species.—Cyprinus tila.

A *Cyprinus* of the *Barilius* kind, devoid of tendrils, with a spot on the middle of each scale upon the sides; with eight rays in the fin of the back; and with ten in that behind the vent.

The *Tila* is found in the northern rivers of Bengal, grows to about a span in length, and is little valued. The *form* is like the head of a lance.

The *head* is sharp and small. The outside of both *jaws* is rough with numerous crowded sharp tubercles. The *mouth* is small, and descends obliquely in a straight line. The *jaws* protrude in opening, and have no *lips*. The *nostrils* are near the *eyes*, which are of moderate size and flat, with circular pupils. Each *gill-cover* contains two plates, and does not conceal its membrane.

The *scales* are large, and adhere firmly. The *shoulder-bones* are naked, sharp, and edged by a membrane. Above each ventral fin is a short scale-like *appendant*.

Each *pectoral* fin has fourteen or fifteen rays. The *ventral* fins do not reach near the vent. The two first of the fins behind the *vent* are very short. The under lobe of the *tail* fin is the longest; and, in all, it contains eighteen distinct rays, besides short compacted ones at each side.

17th Species.—Cyprinus bola.

A *Cyprinus* of the *Barilius* kind, devoid of tendrils, with many oblong diffuse spots, and a longitudinal golden stripe on each side; with ten rays in the fin of the back; and with eleven in that behind the vent.

I found the *Bola (Bhola)* in the Brahmaputra. It grows to four or five inches in length, and is little valued. It is *formed* like the head of a lance. The golden *coloured* line on each side passes between the green and silver parts. The longest diameters of the spots are vertical, so that they resemble imperfect bars. The fins are yellowish.

The *head* is small and sharp. The *mouth* is very large, extending in a straight line farther back than the eyes. The *jaws* do not protrude in opening, and the under one is sharp-pointed and the shortest. The *nostrils* are very near the *eyes*, which are of a moderate size. The *gill-covers* are shining, and conceal part of their *membranes*.

The *scales* are minute. There is only one *lateral line* on each side.

Each *pectoral* fin contains about sixteen rays. The *ventral* fins do not reach the vent. The first ray of the fin behind the *vent* is short, and with the second is undivided: the others are branched. The under lobe of the *tail* fin is the longest, and the fin contains in all nineteen distinct rays, besides short compacted ones.

18th Species.—Cyprinus goha.

A *Cyprinus* of the *Barilius* kind, devoid of tendrils, with many diffuse scattered spots on the sides; with ten rays in the fin of the back; and with thirteen in that behind the vent.

I have found this fish in the Kosi, Yamuna, and Son rivers, where it is called a *Trout* by the English. It grows to about the size of a herring, and is a fine flavoured delicate fish, somewhat resembling in taste the *Smelt* of Europe. In other respects it has the closest resemblance to the *Bola* just now described; and I have only in addition to state, that the spots on the sides of the *Goha* are more numerous, more irregularly scattered, and rounder than those on the *Bola*.

19th Species.—Cyprinus tileo.

A *Cyprinus* of the *Barilius* kind, devoid of tendrils, with many diffuse spots in two rows above each lateral line; with nine rays in the fin of the back; and with fourteen in that behind the vent.

The *Tileo (Tilei)* I found in the Kosi river, where it grows to about a span in length. The *form* is somewhat like that of the blade of a knife, straightish on the back, prominent below, and flat on the sides. The lower fins are of a yellowish *colour*, those of the back and tail dotted.

The *head* is small and bluntish, with a notch at the end of the nose. The *mouth* is large, and descends obliquely. The upper *jaw* protrudes in opening, and is the longer and broader of the two. There are scarcely any *lips*. Each *nostril* has only one opening, near the eye. The *eyes* are far forward. The *gill-covers* are shining, and rounded behind.

The *scales* are of moderate size, and firmly fixed. The *shoulder-blades* are naked and shining. Above each ventral and pectoral fin is a scale-like *appendant*. On each side there is only one *lateral line*.

Each *pectoral* fin contains about fourteen rays. The two innermost rays of each *ventral* fin are closely united. The first ray of the fin behind the *vent* is very short, and is united to the second and third, which, like it, are undivided; the others are branched. The fin of the *tail* has in all nineteen distinct rays, besides short compacted ones.

III. Division.—Cyprinus bangana.

Fishes of the genus *Cyprinus*, with an elevated longitudinal ridge on the middle of the under jaw, as in the genus Mugil, to which these fishes have the utmost resemblance in every thing, except in having no more than one dorsal fin.

The native name *Banggana* is common to most species of the genus *Mugil*, and to those of this division of the genus *Cyprinus*, which perhaps ought to constitute a distinct genus. I have not been able to trace in authors any fish belonging to it. The six last species of this division are so nearly allied to each other, that several of them are probably merely accidental varieties, although this could not be ascertained without a more careful investigation than it was in my power to bestow. All the species of this division, that I have seen, besides the characteristic marks peculiar to the division and genus, have in common the following circumstances:

The *body* is long, and compressed on the sides. The *colour* is greenish on the back, and silver below.

The *dorsal* fin is nearly in the middle of the back, or only a very little behind it, and slopes backward. The last ray is divided to the root. The *pectoral* fins are low down. Each *ventral* fin has nine rays. The fin behind the *vent* is short, containing seven or eight rays.

To this division the 46th species might have been referred; but the ridge on the lower jaw is faintly marked, and its whole appearance is that of the fifth division.

To this division might also have been referred the 83d species; but its ridge also is indistinct, and its appearance is that of the ninth division.

20th Species.—Cyprinus dero. Plate XVII. Fig. 78.

A *Cyprinus* of the *Bangana* kind, with two tendrils; with

thirteen rays in the fin of the back; and with seven in that behind the vent.

The *Dero* (*Dhengro*) I found in the Brahmaputra river. It has the colours and form of the fishes of the seventh division, to be afterwards mentioned, but, on account of the ridge on the lower jaw, must certainly be placed here. It seldom exceeds four inches in length. The *colours* of the back and belly are irregularly indented into each other on the sides. The dorsal and caudal fins are dotted.

The *head* is oval and blunt. The nose projects a little beyond the mouth, and is fleshy, and marked with callous points. At each corner of the mouth is a minute tendril. The *mouth* is small, the upper jaw protruding in opening. The *lips* are fleshy, and smooth on the edges. The *nostrils* are in the middle, between the eyes and point of the nose. The *eyes* are high up, and small, and have circular pupils. The *gill-covers* are almost naked.

The *back* slopes gently before the fin, and is rather sharp-edged. The edge of the *belly* is rounded. The *lateral line* is below the middle, and is bent downward. The *scales* are of a moderate size, and those on the back have some dots.

The second ray of the *back* fin is longer than the first, and shorter than the third,—all the three being undivided, and devoid of joints; the others are branched. The *pectoral* fins are shorter than the head, and each contains about eighteen rays. The *ventral* fins reach to the vent, and are small. The fin behind the *vent* slopes backward, and its first ray is short, and closely united to the second, both being undivided, while the others are branched; and the last is divided to the root. The lobes of the *tail* fin are sharp and equal, and contain nineteen distinct rays, besides several short compacted ones.

21st Species.—CYPRINUS MRIGALA. Plate VI. Fig. 79.

A *Cyprinus* of the *Bangana* kind, with two tendrils; with sixteen rays in the fin of the back; and with eight in that behind the vent.

The *Mrigala* of the Sanscrit, and *Mirgal* of the vulgar dialects, is one of the most beautiful fishes in India, and is pretty common in the ponds and fresh water rivers of the Gangetic provinces. When taken out of the water it is tenacious of life, and grows to from one and a half to two feet in length. It is much valued by the natives, and, although inferior to the *Rohita* and *Catla*, to be afterwards described, is a well tasted fish, of easy digestion.

Its *form* is more prominent on the back than below. The green *colour* on the back changes to golden. The lower fins are tinged red, and all are dotted. The eyes are yellow.

The *head* is bluntish, and half oval, being flat above, and sloping upward below. The *nose* is fleshy, and marked by some hollow dots. On each side it has a very short *tendril*, scarcely reaching to the corner of the *mouth*. This is small, horizontal, and behind the nose. The *jaws* protrude in opening. The *lips* are quite entire on the edges, the under one being erect, and the upper one having no lateral bones. The *nostrils* are in the middle, between the jaw and eye. The *eyes* are globular, and rather small, with circular pupils. The *gill-covers* are rounded behind, and each consists of two plates. Their *membranes* are exposed to view.

The *scales* are dotted and of moderate size. Above each ventral fin there is a large scale-like *appendant*. The *back* forms a high arch, the most elevated part of which is at the commencement of the fin. The *tail* slopes rather suddenly at the fin behind the vent. The *lateral line* runs nearly along the middle of the side, with a very slight bend downwards.

The *dorsal* fin has a concave edge, forming a sharp angle at each end: the first three of its rays are undivided, and closely connected, and the first of them is very short; the other thirteen are branched. The *pectoral* fins are much smaller than the head, and each has seventeen rays, of which the first is undivided. The *ventral* fins are about as large as the pectorals, and do not reach the vent: the first ray of each is undivided, and the others branched. The fin behind the *vent* is near this organ, slopes much backward, and forms an acute angle at the fore end. Its two first rays are undivided, and closely conjoined; the two last are so near that they might, perhaps, be considered as one. The lobes of the *tail* fin are sharp, and nearly equal in length, and have convex edges: in all, the fin has nineteen distinct rays, besides several short compacted ones.

22d Species.—CYPRINUS REBA.

A *Cyprinus* of the *Bangana* kind, with two tendrils; with eleven rays in the fin of the back; with eight in that behind the vent; and with fleshy lips.

The *Reba* is pretty common in the rivers and ponds of Bengal and Behar, especially in their northern parts. It grows to about two feet in length; and, although it is in no great request, does not contain many bones, and has a good flavour. It is subject to very considerable varieties of *colour*. In large clean rivers, its general colour is silver; but toward the back it has some irregular longitudinal dark stripes. In ponds, again, and small rivers full of weeds, the back is green, with a gloss of gold, and the lower fins are tinged with red, which is not the case in large rivers.

The *head* is blunt and oval. The *nose* is very short, and the *tendrils* are near its extremity and minute. The mouth is low, extends horizontally, and is small. The *jaws* protrude in opening; the upper is rather the longest. The *lips* are fleshy. The *nostrils* are near the middle, between the eye and jaw, and each has only one aperture, which is large and circular. The *eyes* are about the middle of the head, large and convex, with the longest diameter of their pupils vertical. Each *gill-cover* contains three plates, and is rounded behind. It exposes to view its *membrane*, which is thick.

The body is covered with adhesive *scales*, which, on the upper parts, are oblong and dotted, and on the belly small. Above each ventral fin is a scale-like *appendant.* The back slopes considerably from the commencement of the fin, and its edge is convex. The *belly* forms an arch. The *lateral line* runs straight along the upper part of the side.

The first ray of the *dorsal* fin is very short, and is closely united with the two next, which, like it, are undivided: the remaining eight are branched. The *pectoral* fins are shorter than the head, with each seventeen rays. The *ventral* fins are of about the same size with the pectorals, do not reach the vent, and do not stand near one another: the first ray of each is undivided. The *anal* fin is near the vent, and slopes backward: its three first rays are undivided, and closely connected, and the others are branched, the last of them being divided to the root. The *tail* fin ends in a crescent, and contains nineteen distinct rays, besides some short compacted ones on each side.

23d Species.—Cyprinus Elanga.

A *Cyprinus* of the *Bangana* kind, with sometimes two tendrils; with nine rays in the fin of the back, and eight in that behind the vent; and with scarcely any lips.

The *Elanga* is another beautiful fish pretty common in the rivers and ponds of Bengal, and in *form* has a considerable resemblance to our *Dace, (Cyprinus Leusiscus.)* On the scales of the sides, towards the back, are little dark lines and dots, forming irregular longitudinal stripes. The fins of the back and tail are blackish. The eyes are green and silver.

The *head* is bluntish, oval, and narrower than the body. In some individuals I have observed a minute *tendril* on each side of the upper jaw; in others I have not been able to trace these organs. The *mouth* descends obliquely with an arch, and is of moderate size. The *jaws* are nearly equal in length, and do not protrude in opening; the upper has a notch at the end. There are scarcely any *lips;* but, in place of the upper, are two small lateral bones. The *tongue* is blunt, smooth, and tied down. The *palate* is smooth. The *nostrils* are near the eyes, which are round and large, with small pupils, the longest diameter of which is vertical. Each *gill-cover* is rounded behind, contains three plates, and exposes to view its *membrane.*

The body is covered with large ill fixed *scales.* The edges of both *back* and *belly* are convex. The *lateral line* descends with a curve parallel to the belly, and is situated below the middle of the side.

The first ray of the *dorsal* fin is short, and closely joined to the second, which, like it, is undivided; the others are branched. The *pectoral* fins are shorter than the head, and each has sixteen rays, of which the first is undivided, and slightly indented on the edge. The *ventral* fins are still shorter, and close to one another. The first ray of each is undivided, and the others branched. The *anal* fin is near the vent, and is nearly triangular: the first of its rays is short, and closely conjoined with the two next, which, like it, are undivided; the others are branched, the last of them being divided to the root. The fin of the *tail* is divided into two lobes, and has nineteen distinct rays, besides short compacted ones.

24th Species.—Cyprinus bata.

A *Cyprinus* of the *Bangana* kind, without tendrils; with twelve rays in the fin of the back, and eight in that behind the vent; with the under lip smooth-edged and erect; with the upper lobe of the tail fin the longest; and with all the fins pale-coloured.

The *Bata* is another beautiful fish, found in the rivers and ponds of Bengal. Its usual size is about a foot in length, but it sometimes grows to eighteen inches. Out of water it is tenacious of life, and spawns in the commencement of the rainy season. In respect of *colour* the fins are much dotted, and that of the tail has an ill defined transverse bar. The eyes are silver, stained with red.

The *head* is oval, and narrower than the body. The *jaws* in opening protrude scarcely any thing; the under one is the shortest. The *lips* are fleshy, the upper having scarcely any lateral bones, and being smooth-edged. The *nostrils* are very near the *eyes*, which are in the middle of the head with respect to its length, are level, and of moderate size, the pupil having its longest diameter vertical. Each *gill-cover* has two plates, and leaves its *membrane* exposed.

The body is covered with large adhesive *scales*, striated on the centre. Above each ventral fin is a scale-like *appendant*. The *belly* is rather more prominent than the back, and the *tail* tapers gradually above and below. The *lateral line* is bent down with a curve.

The first ray of the *dorsal* fin is short, and closely united with the second, which, like it, is undivided; the others are branched, the last of them being divided to the root. The *pectoral* fins are low, and are nearly as long as the head; each has seventeen rays, of which the first is undivided. The *ventral* fins are smaller than the pectorals: the first and last rays of each are undivided. The *anal* fin is near the vent, and slopes behind: the first of its rays is very short, and is closely joined to the two following, which, like it, are undivided; the others are branched, and the last of them is divided to the root. The *tail* fin contains nineteen distinct rays, besides short compacted ones.

25th Species.—Cyprinus acra.

A *Cyprinus* of the *Bangana* kind, without tendrils; with eleven rays in the fin of the back, and eight in that behind the vent; with the under lip smooth-edged, and erect; with the upper lobe of the tail fin the longest; and with the lower fins reddish.

This fish, which I found in the Sunkos river, in the north-eastern part of Bengal, has the utmost resemblance to the *Bata*, so that I have only noted the few points in which it differs from the description above given; and, besides what is mentioned in the specific character, these are as follows: Each *pectoral* fin has eighteen or nineteen rays. The *scales* are scarcely striated on the middle, and near the edge are very smooth.

26th Species.—Cyprinus cura.

A *Cyprinus* of the *Bangana* kind, without tendrils; with twelve rays in the fin of the back, and seven in that behind the vent; with the under lip smooth-edged, and erect; and with the lobes of the tail fin equal in length.

The *Cura* is another fish so nearly allied to the *Bata*, that I have only noted the circumstances in which it differs from the account before given. It is found in the rivers of Bengal, and scarcely exceeds the finger in length. The fins are of a yellowish *colour*, and very little dotted. In the fin behind the *vent* the two first rays only are undivided. On each *gill-cover* is a very narrow crescent-formed mark, indented on the edges.

27th Species.—CYPRINUS PANGUSIA.

A *Cyprinus* of the *Bangana* kind, without tendrils; with fourteen rays in the fin of the back, and eight in that behind the vent; with the under lip reflected, and entire on the edges; and with large pores on the nose.

The *Pangusia* (*Pangusiya*) I found in the Kosi river, where it grows to about a span in length. In *colour* it has no appearance of longitudinal stripes, but the edges of the scales are dotted. The eyes are silver, stained with red.

The *head* is oval, and narrower than the body. The *jaws* protrude in opening. The *lips* are fleshy, and both are smooth on the edges. The *nose* is fleshy. The *nostrils* are very near the eye. The *eyes* are near the middle of the head, in respect to length are of moderate size, and the longest diameters of their pupils are vertical. The *gill-covers* are rounded behind, and do not conceal their *membranes.*

The body is covered with large *scales*, adhering firmly, and even on the surface. Above each ventral fin is a scale-like *appendant.* The *tail* tapers gradually. The *lateral line* descends with a curve along the middle of the side.

The *dorsal* fin is dotted: the two first of its rays are very short, and, with the two following, are undivided; the last ten are branched. The *pectoral* fins are shorter than the head, each having seventeen rays. The *ventral* fins do not reach the *vent.* The fin behind this organ slopes backward, and the three first of its rays are undivided, and lengthen gradually from the first; the five last are branched, and the last of them is divided to the bottom. The fin of the *tail* contains nineteen distinct rays, besides compacted ones, and is divided into two lobes, of which the under one is the shortest and bluntest.

28th Species.—Cyprinus ariza. Buchanan's Journey through Mysore, &c. Vol. III. p. 344, Plate XXXI.

A *Cyprinus* of the *Bangana* kind, without tendrils; with twelve rays in the fin of the back, and seven in that behind the vent; with the under lip smooth on the edges, and reflected; and without large pores on the nose.

The *Ariza* is found in the rivers of Bengal as well as in those of the peninsula of India, where I first saw it. There is the utmost resemblance between it and the *Pangusia* last described, and both grow to about the same size. I shall only mention the circumstances, besides those in the specific character, by which it differs from the description of the *Pangusia* before given.

The *belly* is more prominent than the back. In the *dorsal* fin three only of the rays are undivided. In each *pectoral* fin there are sixteen rays. In the fin behind the *vent* there are only two undivided rays. The lower fins are yellow, or slightly stained with red.

29th Species.—Cyprinus boga. Plate XXVIII. Fig. 80.

A *Cyprinus* of the *Bangana* kind, without tendrils; with twelve rays in the back fin, and eight in that behind the vent; with the under lip indented on the edge; and with large pores on the nose.

The *Boga* is another fish so nearly resembling the two preceding, that, besides the circumstances mentioned in the specific character, I shall mention only those by which it is chiefly distinguished. I have found it in the Brahmaputra, where it grows to about a foot in length.

The *eyes* are redder than in either of the two last species. Each *scale* along the middle is marked with a slender longitudinal ridge; but the ridges forming the proper *lateral lines* are much larger than those on the scales above or below. The *back* is more prominent than the belly. The fin of the *back* slopes behind with a considerable curve, and can be thrown back in a deep furrow. It contains three undivided rays, as does also the fin behind the *vent.* The fin of the *tail* is not so deeply divided as in the other two species, and ends nearly in a crescent. All the fins, except the pectorals, are more or less stained with black.

IV. Division.—Cyprinus, properly so called.

Fishes of the *Cyprinus* genus, without any affinity to another genus; of a large size and thick form; devoid of spots; and with the lateral lines near the middle.

The fishes of this division, although forming a natural group, have no generic appellation among the natives; and, being the best defined species, I have considered them as the prototypes of the genus, and have therefore given the division no peculiar name.

Besides the marks common to the genus, and those characteristic of the division, the only circumstance common to these fishes, that I observe, is, that they all have the fin behind the vent very short.

30th Species.—Cyprinus catla. Plate XIII. Fig. 81.

A *Cyprinus* proper, without tendrils; with eighteen rays in the fin of the back, and eight in that behind the vent; with the body of a dirty silver colour, having some dusky green on the back; and with the lips smooth-edged.

The *Catla* is a common fish in the rivers and tanks of Bengal, but is seldom found farther west, and even in most parts of Behar is unknown. It grows to about three or four feet in length, and is a light wholesome fish, free of bones. The head is very fat and delicious, and the flavour of the whole is excellent, when the size is moderate, that is, when the length is from eighteen to twenty-four inches; but when large it acquires a rank taste. It is a very strong active animal, and often leaps over the seine of the fishermen, on which account, when fishing for the *Catla*, they usually follow the net in canoes, and make a noise by shouting and splashing with their paddles.

In form, quality, and habits, this fish has a considerable resemblance to the *Carp* of Europe, (*Cyprinus Carpio*,) but it wants the tendrils on the jaws, with which that fish is provided. The *Catla* is a deep full *formed* fish, but a good deal compressed. The fins are darkish *coloured*, and no part except the eyes have any tinge of red.

The *head* is large, oval, blunt, a little wider than the body, and smooth. The *mouth* is of a moderate size, and descends obliquely from the extremity of the nose. The *jaws* are nearly equal in length, and protrude in opening. The *lips* are smooth and fleshy; the upper has no bones, and the under is reflected. The *palate* is smooth. The *tongue* is blunt, undivided, fleshy, smooth, and tied down through its whole length. The *nostrils* are high up and very near the *eye*, which is convex and of a good size, and has a circular pupil. Each *gill-cover* is rounded behind, and edged by a broad membrane, and consists of three plates. The bones supporting the *gills* have on the inside a row of slender teeth.

The *scales* are large and entire on the edge, and each is separated into two divisions by a dotted vertical line. The *back* is more prominent than the belly, and arched, rising before the fin with a slope greater than that of the head. The *tail* slopes rather suddenly at the fin behind the vent. The *lateral line* forms a curve with the concavity upward.

The *dorsal* fin is near the middle of the body, and slopes backward with a concave edge. The first of its rays is short, and closely joined to the second, which, like it, is undivided and smooth : the others are branched, and the last is divided to the root. The *pectoral* fins are much smaller than the head. The *ventral* fins are of about the same size with the pectorals, reach beyond the vent, and stand at a little distance from one another. The two first rays of each are undivided, the first being very short. The fin behind the *vent* is in the middle of the tail, and slopes much backward ; the three first of its rays are undivided, and situated close to each other, and lengthen gradually from the first to the third ; the others are branched, and the last of them is divided to the root. The fin of the *tail* is divided beyond the middle into two sharp equal lobes, and contains nineteen distinct rays, besides several short compacted ones on each side.

31st Species.—Cyprinus curchius.

A proper *Cyprinus*, without tendrils ; with fifteen rays in the fin of the back, and seven in that behind the vent ; and with very small scales.

The *Curchius (Kurchi)* is found in the fresh waters of Bengal, and is a beautiful fish, from one foot to a foot and a half in length ; but is of little value, as it is full of bones. It is of an oblong compressed *form*, nearly like the head of a lance, but rather more prominent below than above. The *colour* is green above and silver below, with fins inclining to olive, and blackish eyes. The dorsal and anal fins are dotted with black, the dots on the latter forming a stripe along the middle of the fin.

The *head* is oval, and of moderate size. The *nose* is dotted, and projects a little beyond the *mouth*, which is of a moderate size. The *jaws* protrude in opening. The *lips* are fleshy, pendulous, and smooth-edged. The *nostrils* are high, and far back towards the eyes. The *eyes* are high, convex, and of moderate size, with oval pupils, their longest diameters being vertical. The *gill-covers* are rounded behind, and smooth-edged, and do not conceal their *membranes*.

Above the ventral fins there is scarcely any scale-like *appendant*. The *lateral line* commences at the upper end of the gill-cover, and is bent slightly down. The *vent* is behind the middle.

The *dorsal* fin occupies nearly the middle of the back, and slopes backward: the first of its rays is short, slender, and closely conjoined to the second, which is also undivided: the others are branched, and the last of them is divided to the root. The *pectoral* fins are much shorter than the head: each has eighteen rays. The *ventral* fins are at some distance from each other, and about as large as the pectorals; each contains nine rays. The fin behind the *vent*, on the middle of the tail, slopes backward: the first of its rays is short, and adheres to the second, which, like the first, is undivided: the others are branched, and the last is divided to the root. The fin of the *tail* is divided into two lobes, and contains nineteen distinct rays, besides some short compacted ones on each side.

32d Species.—CYPRINUS CURSA.

A proper *Cyprinus*, with two tendrils; with sixteen rays in the fin of the back, and eight in that behind the vent; and with small scales.

This, with the preceding, and two following species, on account of their small scales and size, have a considerable affinity to the *Tench* of Europe, *(Cyprinus tinca;)* but they are inferior fishes for the table, and all of them have a prominent belly; while that of the *Tench* is nearly straight. The *Cursa*, having a tendril at each corner of the mouth, has the greatest resemblance to the *Tench*, but is an abundantly distinct species. The *Cursa* has also a considerable resemblance to the *Cyprinus capoeta*, (Bonnaterre, *Tabl. Encycl.* p. 191, Pl. C. Fig. 411;) but that fish has ten rays in each ventral fin.

The *Cursa* inhabits fresh water rivers and ponds in the south of Bengal, and is often found from two to three feet in length. It is full of bones, and many of the natives abstain from its use, imagining, that, if eaten in the same day with milk, it will occasion the disease called *elephantiasis.* It is of an oblong compressed *form*, and silver *colour*, with a dusky back. All the fins, except the pectorals, are blackish, with numerous dots.

The *head* is oval, blunt, and small. The *tendrils* are minute. The *nose* projects a little beyond the *mouth*, is fleshy, and perforated with a good many callous pores. The *mouth* is small and straight. The *lips* are fleshy and pendulous, the upper one having no bones. The *nostrils* are in the middle between the jaw and eye. The *eyes* are high, and of moderate size, with the longest diameter of the pupil vertical. Each *gill-cover* is rounded behind, consists of three plates, and exposes its membrane.

Above each ventral fin is a small scale-like *appendant.* The *back* slopes down to the head. The *belly* is arched. The *lateral line* is rather high, and is bent down in the middle. The *vent* is behind the middle.

The *dorsal* fin occupies the middle of the back, and slopes backward with a hollow edge: the three first of its rays are undivided, and the first is very short; the others are branched, and the last of them divided to the root. The *pectoral* fins are low, sharp above, and smaller than the head: each has seventeen rays, of which the first is undivided. The *ventral* fins are smaller than the pectorals, and separated from each other: each has nine rays, of which the first is undivided. The fin behind the *vent* is before the middle of the tail, and slopes backward, with a concave edge: the three first of its rays are undivided, and closely conjoined, the first being very short; the others are branched, and the last of them divided to the root. The fin of the *tail* is divided into two lobes, and contains nineteen distinct rays, besides short compacted ones at each side.

33d Species.—CYPRINUS CURSIS.

A proper *Cyprinus*, with four tendrils; with sixteen rays in the fin of the back; and eight in that behind the vent; and with small scales.

The *Cursis* (*Kursi*) I found in the river Kosi of Northern Behar. Except in the few circumstances mentioned below, and in the specific character, it agrees entirely with the description of the *Cursa;* and I have some doubt of the species being actually different.

The *colour* is silver, with black dots disposed on the back in ill defined longitudinal stripes. On the sides are scattered many red atoms. The *lips* are toothletted or fringed; the under one is double.

34th Species.—CYPRINUS GONIUS. Plate IV. Fig. 82.

A proper *Cyprinus*, with four tendrils; with fifteen rays in the fin of the back, and seven in that behind the vent; with large pores on the nose; with fringed lips; with longitudinal lines of black dots on the sides; and with moderate sized scales.

The *Gonius (Goni)* is a strong fish, very tenacious of life, but little valued for food. It grows to a foot and a half in length, and is found in the fresh water rivers and ponds of Bengal.

The *form* is like the head of a lance, compressed at the sides, and rather more prominent on the back than below. Above it is of a dark green *colour* with a gloss of silver; below the colour is entirely silver, and on the sides are several longitudinal lines composed of black dots. The fins are darkish, and all of them dotted with black. The eyes are silver with a tinge of green.

The *head* is oval, small, and blunt. The *nose* projects a little beyond the mouth, and is marked by some large hollow dots. The *mouth* is small. The *jaws* protrude in opening. The *lips* are fleshy, pendulous, and indented, or fringed, on the edges: the upper one has no bones. The *tendrils*, which are very short, proceed from the corners of the lips. The *nostrils* are nearly in the middle between the eyes and jaw. The *eyes* are high, of moderate size, and have the longest diameter of their pupils vertical. The *gill-covers* are rounded behind, and edged with a broad membrane: they do not conceal their proper *membranes*, the rays in which are flat.

The *scales* are roughish, and of a moderate size. Above each ventral fin is a scale-like *appendant* covered with small scales. The *lateral lines* are ill defined, and descend with a curve along the middle of the sides. The *vent* is behind the middle.

The *dorsal* fin occupies the middle of the fish, and slopes backward with a concave edge, ending very sharp behind. The two first of its rays are undivided and conjoined, the first being a half shorter than the second, which is smooth; the others are branched. The *pectoral* fins are low, much smaller than the head, and sharp above: and each contains about seventeen rays. The *ventral* fins are large, but do not reach to the vent: and each has nine rays. The fin behind the *vent* is before the middle of the tail, and slopes backward. The two first of its rays are strong smooth undivided bones, closely joined to each other, the second being twice as long as the first: the others are branched, and the last of them is divided to the root. The *tail* fin is slightly divided into two lobes, or rather ends almost in a crescent.

35th Species.—CYPRINUS CURMUCA. Buchanan's Journey to Mysore, &c. Vol. III. p. 334, Plate XXX.

A proper *Cyprinus*, with two tendrils; with eleven rays in the fin of the back, and eight in that behind the vent; with moderate sized scales; with callous tubercles on the head; and with the third ray of the dorsal fin smooth behind.

This and the following species have a very great resemblance to the *Cyprinus cirrhosus* of Bloch, (*Ichth.* Tome XII. p. 43, Pl. CCCCXI.;) but the callosities of the head are neither mentioned in Bloch's description, nor represented in his figure.

The *Curmuca* is found in the fresh water rivers of the south of India, and sometimes reaches three feet in length. It is of an oblong compressed *form*, shaped like the head of a lance, and is of a dusky green above and of a silver *colour* below.

The *head* is small and oval. Near the upper jaw are scattered many small callous projecting tubercles. The *tendrils* are minute, one at each corner of the *mouth*, which is small, and a little below the extremity of the nose, and extends straight backwards. The *jaws* protrude much in opening; the under one is the shortest. The *lips* are smooth-edged, and the under one is erect, while the upper has a small bone on each side. There is scarcely any *tongue*. Across the middle of the *palate* is suspended a membranous veil. The *nostrils* are near the *eyes*, which are high, of a good size, flat, and round, with circular pupils. The *gill-covers* are rounded behind, and each contains three bony plates, which do not conceal their *membranes*.

The body is covered with moderate sized *scales*, which adhere firmly. Above each ventral fin is a scale-like *appendant.* The *lateral line* descends a little at the shoulder, and then runs straight along the middle of the side. The *tail* slopes gradually. The *vent* is behind the middle.

The *dorsal* fin is nearly in the middle, and slopes much backward. The three first of its rays are smooth, closely conjoined, and undivided; and the first is very short: the others are branched, and the last of them divided to the root. The *pectoral* fins are much smaller than the head, and sharp above; and each contains sixteen rays, of which the first is undivided. The *ventral* fins are larger than the pectorals, extend beyond the vent, and are at some distance from one another; and each contains nine rays, of which the first is undivided. The fin behind the *vent* is on the middle of the tail, and is rather rounded. The first of its rays is very short, and closely united with the two next, which, like it, are undivided; the others are branched, and the last of them is divided to the root. The tail *fin* consists of two sharp lobes, of which the upper is rather the longest and sharpest: both contain eighteen distinct rays, besides some short compacted ones.

36th Species.—Cyprinus chagunio.

A proper *Cyprinus,* with four tendrils; with twelve rays in the fin of the back, and eight in that behind the vent; with callous tubercles on the head; with large scales; and with the third ray of the dorsal fin indented behind.

The *Chagunio (Chaguni)* I have found in the Yamuna, and in the northern rivers of Behar and Bengal. It grows to about a foot and a half in length, and is considered good eating. It is rather deep in *form,* and its back is more prominent than the belly. The *colour* is green above, and below silver; and the scales towards the back, being dotted at the base, give an appearance of longitudinal stripes.

The *head* is narrower than the body, and small. The tubercles are round, and scattered over the fore part of the head. The *tendrils* are minute, two proceeding from the sides of the upper jaw, and two from the corners of the mouth. The *mouth* is low and small, and extends straight backward. The *jaws* protrude in opening; the upper is the longest. The *lips* are fleshy, and smooth-edged. The *tongue* is smooth, sharp, thick, and tied down throughout its whole length. From the middle of the *palate* is suspended a loose membrane. The *nostrils* are near the *eyes*, which are high, flat, large, and round, with circular pupils. The *gill-covers* are rounded behind, and do not conceal their *membranes*, the bones in which are flat.

The *back* slopes gradually in each direction from the beginning of the fin, and its edge forms a sharp ridge. The *belly* is almost straight. The *tail* tapers gradually. The *lateral line* descends a little, and is rather below the middle of the side. The *vent* is behind the middle. The body is covered with large smooth-edged *scales*, adhering firmly. Above each ventral fin is a scale-like *appendant*.

The *dorsal* fin occupies the middle of the back, and slopes behind. The three first of its rays are strong undivided bones, the teeth in the last of them being long and reflected; the other nine rays are branched, and the last of them is divided to the root. The *pectoral* fins are low, much shorter than the head, and sharp above. Each contains seventeen rays, of which the first is undivided. The *ventral* fins are at some distance from each other, and are about as large as the pectorals, each having ten rays, of which the first is very small, and the second undivided. The fin behind the *vent* occupies the middle of the tail, and slopes backward: the three first of its rays are closely united and undivided; and the others are branched, the last of them being divided to the root. The fin of the *tail* is divided into two equal lobes, and contains nineteen distinct rays, besides several short compacted ones.

37th Species.—CYPRINUS CALBASU. Plate II. Fig. 83.

A proper *Cyprinus*, with four tendrils; with sixteen rays in the fin of the back, and eight in that behind the vent; with the nose smooth, and projecting considerably beyond the mouth; and with lips indented on the edge.

The *Calbasu* (*Kalbosu*) is very common in the rivers and ponds of Bengal, and is also found in the western provinces. It has a strong affinity to the *Barbell* (*Cyprinus barbus*) of England, but is deeper in the form. It is often found a foot and a half in length, and sometimes twice that size, and is a well tasted light food, but contains many small bones. It is of an oblong compressed *form*, more protuberant above than below, and, in general, is of a dark olive *colour*, with a silvery gloss, inclining to yellow, on the belly, while the scales are covered with numerous black dots. In some places, as at Mungger, where the water is clear, on a rocky or pure sandy bottom, many of the scales on the sides assume a ferruginous colour, and the fish is then considered as of a distinct species, and called *Kundhna;* but I have no doubt that this difference of colour is a mere accidental circumstance. The eyes are orange-coloured.

The *head* is blunt, small, and devoid of tubercles or large pores. The *tendrils* are short, thick, and incurved, the two longest proceeding from the end of the nose, and the two shorter from the middle of the upper jaw. The *nose* projects considerably beyond the mouth, and is covered with minute dots. The *mouth* descends obliquely, and is small. The *jaws* do not protrude in opening. The *lips* are fleshy, pendulous, and indented on the edges, the upper one having very minute, or rather no bones. The *nostrils* are far back, and each has only one aperture, covered by a membranous valve. The *eyes* are high, and rather small, with circular pupils. The *gill-covers* are rounded behind, smooth on the edge, contain each three plates, and conceal in part their *membranes.*

The ridge of the *back* forms an arch, rising before more suddenly than the head, and terminating near the end of the tail. The fore part of the arch is blunt, the hinder part has a sharp edge. The *belly* is nearly straight, but the tail contracts suddenly at the fin behind the vent. The *lateral line* descends with a slight curve from the shoulder along the middle of the side. The *vent* is behind the middle. The body is covered with large, dotted, and firmly-fixed *scales,* without any scale-like *appendants* above the ventral fins.

The *dorsal* fin occupies the middle of the back, and slopes only a little behind. The first of its rays is a half shorter than the second, to which it closely adheres, and, like it, is undivided: the others are branched. The *pectoral* fins are low, shorter than the head, and sharp above, and each contains eighteen rays. The *ventral* fins are near each other, and are about as large as the pectorals: each has nine rays, of which the two first are undivided. The fin behind the *vent* occupies the middle of the tail, and slopes much backward; the two first of its rays resemble those in the dorsal fin; the others are branched. The fin of the *tail* is divided into two bluntish lobes, of which the upper is rather the longest. In both are nineteen distinct rays, besides some short compacted ones.

38th Species.—Cyprinus nancar.

A proper *Cyprinus*, with four minute tendrils; with twenty rays in the dorsal fin, and eight in that behind the vent; with the edges of the lips smooth; with no tubercles nor large pores on the nose; and with large equal scales on the sides.

The *Nancar* has a considerable affinity to the *Cyprinus gibellio* of Bloch, (*Ichth.* Tome I. p. 61, Pl. XII.) which, however, wants the tendrils. I found it in the small rivers of the Gorakhpur district, such as the Gunggi. It does not attain a greater weight than three pounds, and is of a deep compressed *form*, above of a dark green *colour*, with a golden gloss, and below white. The fins are dark coloured, and the eyes reddish.

The *head* is small, blunt, and oval. The *tendrils* proceed from the corners of the mouth. The *nose* projects a little beyond the mouth, and is fleshy and bluntish. The *mouth* is low, extends straight back, and is small. The *jaws* protrude in opening, and are nearly equal in length; within the upper one a pendulous membrane is suspended. The *lips* are fleshy, and the under one is erect. The *eyes* are high, circular, and of moderate size. Each *gill-cover* consists of one bony plate, rounded behind, and does not conceal its *membrane*.

The ridge of the *back*, towards the fin, is sharp-edged. The *belly* forms an arch. The *lateral line* descends with a curve. The *vent* is behind the middle of the body. The *scales* are large, adhere firmly, and are minutely dotted. Above each ventral fin there is a very short scale-like *appendant*.

The *dorsal* fin is behind the middle. The first of its rays is very short, and, with the second and third, is undivided: the others are branched, and the last of them divided to the root. The *pectoral* fins are smaller than the head, and low, and each contains about eighteen rays. The *ventral* fins do not reach the vent, and each has nine rays. The three first rays of the fin behind the *vent* are closely united, and undivided. The *tail* fin consists of two sharp lobes, and contains about twenty distinct rays, besides some short compacted ones.

39th Species.—CYPRINUS NANDINA. Plate VIII. Fig. 84.

A proper *Cyprinus*, with four tendrils; with twenty-six rays in the fin of the back, and eight in that behind the vent; with the lips fringed, and the nose smooth; and with large scales.

The *Nandina* (*Nandin*) I found in the Mahananda river, and in the large adjacent marshes or lakes which surround the ruins of ancient Gaur. In the Gorakhpur district I found a fish of the same name, which differed in no respect from the description given below, except that it had only twenty-three rays in the dorsal fin; and I doubt much if this variation can be considered sufficient to constitute a distinct species. The *Nandina*, of all the fishes that I have seen in India, has the greatest affinity, in external appearance, to the *Carp* (*Cyprinus carpio*) of Europe; but many of its qualities are very different. Of all Bloch's fishes, the *Cyprinus fimbriatus* (*Ichth.* Tome XII. Pl. CCCCIX.) has the greatest resemblance to the *Nandina*: but its sides are spotted with red, and, according to La Cepede, (*Hist. des Poissons,* Tome V. p. 502,) it has no tendrils; and it has tubercles on the nose, and only eighteen rays in the dorsal fin.

The *Nandina* often grows to two, or even three feet in length, and is a well-tasted fish. Its *form* is deep and thick; but still it is compressed. It is of a dark green *colour* above, with a coppery gloss, and white below, with dark fins and red eyes.

The *head* is small, blunt, and half oval, being rather flat above. From each corner of the mouth there are two minute *tendrils*. The *nose* is blunt and fleshy, without tubercles, or remarkable pores. The *mouth*, which is low and small, opens straight back. The *jaws* protrude in opening, and each has within a transverse membrane, fixed by one edge. The *lips* are fleshy, and edged with many short sharp slender tendrils: the under one is reflected. The *nostrils* are nearly in the middle, between the eyes and nose.

Although both form arches, the *back* is rather more prominent than the *belly:* towards the fin the former has a sharp edge. The *lateral line* descends with a curve from the shoulder. The *vent* is behind the middle. The *scales* are large, and marked with dots and concentric circles. Above each ventral fin is a scale-like *appendant.*

The *dorsal* fin is rather behind the middle, and contains twenty-six rays, or, in one variety, only twenty-three. The first three rays are closely conjoined, and undivided, the first being very short: the others are branched; and the last divided to the root. The *pectoral* fins are much smaller than the head, low placed, and sharp above, each containing sixteen rays. The *ventral* fins are about as large as the pectorals, and do not reach the vent: each has nine rays. The fin behind the *vent* slopes much backward, and the three first of its rays are undivided, and closely united, the first being very short; the others are branched, and the last divided to the root. The fin of the *tail* is deeply divided into two sharp lobes, and in all contains twenty-three rays.

40th Species.—CYPRINUS ROHITA. Plate XXXVI. Fig. 85.

A proper *Cyprinus,* with two tendrils; with fifteen rays in the back fin, and seven in that behind the vent; with large scales; with the nose smooth and devoid of remarkable pores; with the lips indented on the edges; and with the under one reflected.

The *Rohita,* in its form, has a considerable resemblance to the *Cyprinus nasus,* (La Cepede, *Hist. des Poissons,* Tome V. p. 585;) but, in its size and qualities, has a stronger resemblance to the *Carp, (Cyprinus carpio,)* which, however, it appears to me to excel; and it is not only a fish of easy digestion, and well flavoured, but the belly is uncommonly rich and delicious. It abounds in the fresh water rivers of the Gangetic provinces, and also in those of Ava, and, in the ponds of the former, is propagated with considerable care, being considered as the most valuable fish; although I think, that what is found in the smaller and more rapid rivers is by far the best for eating. Its beauty, both in form and colour, equals its value for the table; and of all fresh water fishes, so far as I have seen, it is, perhaps, the most excellent. It frequently grows to three feet in length, and is of a long but full made *form*, rather more prominent above than below. The tail contracts suddenly at the fin behind the vent. The back is of a green *colour*, and the belly silver; but on the sides the middle of the scales is golden. The eyes and all the fins, except the dorsal, are reddish.

The *head* is blunt, smooth, and half oval, sloping much below and little above. It is small, but rather wider than the body. The *nose* is fleshy and blunt, is devoid of tubercles or remarkable pores, and projects some way beyond the *mouth*, which is small, and extends straight backward. At each of its corners is a minute *tendril.* The *jaws* protrude in opening, the upper being the longest, and having a membrane stretched across its inside. The *lips* are fleshy, and indented on the edges with short thick processes. The *tongue* is tied down throughout its whole length. The *nostrils* are nearer the eyes than the nose. The *eyes* are high, and of a good size, and have circular pupils. Each *gill-cover* consists of three plates, is rounded behind, and does not conceal its *membrane,* which is thick, with flat rays. The bones supporting the rays are unarmed, and their inside forms a ridge.

The *back* forms a regular arch to near the end of the tail.

The *belly* is almost straight. The *lateral line* descends, with a slight curve, from the shoulder. The *vent* is behind the middle. The *scales* are large, and dotted round the edges. Above each ventral fin is an *appendant*, consisting of two scale-like plates.

The *dorsal* fin occupies the middle of the back, and slopes behind. The first of its rays is very short, and is closely conjoined with the two next, which, like it, are undivided: the others are branched, and the last is divided to the root. The *pectoral* fins are low, much shorter than the head, and sharp above: each has eighteen rays, of which the first is undivided. The *ventral* fins are about as long as the pectorals, are placed at some distance from each other, and each has ten rays, of which the first is very short, and closely united to the second, which is undivided. The *anal* fin is in the middle of the tail, and slopes much behind: the first of its rays is short, and closely united to the second, which is undivided, the others are branched, and the last is divided to the root. The fin of the *tail* is divided into two lobes, nearly equal, and bluntish, and has nineteen distinct rays, besides short compacted ones.

41st Species.—CYPRINUS PUTITORA.

A proper *Cyprinus*, with four tendrils; with eleven rays in the fin of the back, of which two are undivided; with seven rays in the fin behind the vent; with very large scales; with a smooth nose; with smooth-edged lips; and with pale fins.

This and the two following species have, in many points, a strong resemblance, being very large fishes, affording an excellent wholesome nourishment, free from bones, although not quite so delicate as the *Rohita.* They are all also strong, well-formed, handsome fishes, peculiarly distinguished by the enormous size of their scales, which, in large individuals, almost equals the hand, insomuch, that cards for gaming are sometimes made of them at Dakha. *Mahasaula* and *Tora*, variously altered or corrupted, or with various additions, may be considered as generic appellations among the natives for these fishes, all of which frequent large rivers.

The *Putitora* is found in the eastern parts of Bengal, and sometimes grows to nine feet in length. It is of an oblong compressed *form*, with, however, a great width; and above is of a dusky *colour* with a gloss of steel, while the edges of the scales change from gold to silver. The lower parts resemble entirely the latter. The fins are without spots, and the hinder ones are tinged with yellow. The eyes are like silver.

The *head* is blunt, oval, small, and smooth, with a very minute *tendril* at each corner of the mouth, and another from each side of the upper jaw. The *mouth* is small, and opens horizontally. The *jaws* protrude in opening, and the upper is the longest. The *lips* are fleshy, and the under one is reflected. The *nostrils* are in the middle between the eyes and nose. The *eyes* are far forward, and small, with circular pupils. The *gill-covers* have little or no flesh, and do not conceal their *membranes*.

The *back* from the fin to the head slopes and forms a sharp ridge. The *tail* is suddenly narrowed at the fin behind the *vent*, which is behind the middle. The *lateral line* is scarcely distinguishable. The *scales* end with a wide notch. Above each ventral fin is a scale-like *appendant*.

The *dorsal* fin occupies nearly the middle of the back, and slopes behind: the first of its rays is very short, and the second very strong and undivided; the others are branched, and the last of them divided to the root. The *pectoral* fins are low, shorter than the head, and sharp above; each contains fifteen rays. The *ventral* fins do not reach the vent; each contains nine rays. The fin behind the *vent* is rounded, and may be almost entirely withdrawn into a scaly furrow. The first of its rays is very short, and closely united to the second, which is undivided; the others are branched, and the last of them divided to the root. The fin of the *tail* is divided into two lobes, of which the under is the sharpest. Both contain nineteen distinct rays, besides short compacted ones.

In one specimen I found only two tendrils, with ten rays in the dorsal fin, and eight in that behind the vent; but, as it differed in no other particulars, and was considered by the fishermen as of the same species, I suppose that these circumstances were owing to accident.

42d Species.—Cyprinus tor.

A proper *Cyprinus*, with eleven rays in the fin of the back, (of which three are undivided,) and eight in that behind the vent; with very large scales; with a smooth nose, and smooth-edged lips; and with reddish fins on the belly.

This fish I found in the Mahananda river, where it grows to three or four feet in length. It is of a long compressed *form*, but considerable width, rather more prominent below than above. Above its *colour* shines with gold and green, below like silver. The fins have no spots, but the dorsal one is dotted. The eyes are silver-coloured, with some dots above.

The *head* is sharpish, half oval, (being flattish above,) and smooth, with a protuberance between the nostrils, arising from a process of the upper jaw, which passes backward under the skin. There is a minute *tendril* at each corner of the mouth, and two at the extremity of the nose. The *mouth* is low, and extends straight backward. The *jaws* protrude in opening. The *lips* are fleshy, and the lower terminates in a narrow reflected point. The *nostrils* are near the eyes. The *eyes* are high, of moderate size, and have circular pupils. The *gill-covers* are not bordered behind by a membrane, but expose their proper *membranes* to view.

The *back* slopes from the fin to the head, and has a blunt or convex edge. The *vent* is behind the middle. The *lateral lines* bend down from the shoulder. The *scales* adhere firmly, and are dotted at the edges. Above each ventral fin is a scale-like *appendant.*

The *dorsal* fin occupies the middle of the back: the three first of its rays are very strong, undivided, and closely united, the first being very short, and the third the longest; the others are branched, and the last is divided to the root. The *pectoral* fins are shorter than the head, and sharp above: each has eighteen rays, of which the first is undivided. The *ventral* fins do not extend to the vent, and each has nine rays. The *anal* fin is near the vent: the three first of its rays are undivided, and gradually lengthen from the first, which is very short; the other five are branched, and divided to the root. The *tail* fin is divided into two lobes, and has nineteen distinct rays, besides short compacted ones.

43d Species.—Cyprinus mosal.

A proper *Cyprinus*, with four tendrils; with thirteen rays (of which the four first are undivided) in the fin of the back, and eight in that behind the vent; with very large scales; with a smooth nose, and smooth-edged lips; and with pale fins.

The *Mosal* I found in the river Kosi, where it grows to four or five feet long. It is of a long, compressed, but thick *form*, more prominent on the back than on the belly, its *colours* shining above with green and gold, and below with silver. The fins have no spots, and a slight orange tinge. The eyes are silver, dotted above.

The *head* is sharp, oval, and narrower than the body, and

smooth, with a small tubercle between the nostrils, as in the last species, but less strongly marked. There is a minute *tendril* from each corner of the mouth, and two from the middle of the nose. The *mouth* is low, and extends straight back. The *jaws* protrude in opening, the under one being shortest. Both the *lips* are fleshy and rounded, nor is the under one reflected. The *nostrils* are near the eyes. The *eyes* are high, and of moderate size. The *gill-covers* are not margined behind with a membrane, but expose to view their proper *membranes*.

The *back* slopes from the fin forward, and there forms a sharpish ridge. The *vent* is behind the middle. The *lateral line* bends down from the shoulder. The *scales* adhere firmly, and are rounded, those of the lateral line terminating in a small notch. Above each ventral fin is a scale-like *appendant*.

The *dorsal* fin occupies the middle of the back; the three first of its rays are small, and the fourth is a very large undivided bone; the others are branched, and the last of them divided to the root. The *pectoral* fins are low, shorter than the head, and sharp above, each having about seventeen rays. The *ventral* fins do not extend to the vent, and each has nine rays. The *anal* fin is near the vent; the three first of its rays are undivided, the second being longer than the first, and shorter than the third; the others are branched, and the last divided to the root. The *tail* fin is divided into two lobes, and has nineteen distinct rays, besides short compacted ones.

44th Species.—Cyprinus sarana.

A proper *Cyprinus*, with four tendrils; with ten rays in the fin of the back, the longest of the rays being indented behind; with eight rays in the fin behind the vent; with large scales;

with the jaws nearly equal; with the under lip erect, and smooth-edged; and with pale fins.

I have little or no doubt that this is the same with the *Kunamoo* of Dr Russell, (Indian Fishes, No. 24,) although he mentions two small bony rays before the largest one, while I have noticed only one; and, although he reckons nine rays in the anal fin, while I reckon eight, he considering as two distinct rays what I call one divided to the root.

The *Sarana* would also appear to have a considerable affinity with the *Cyprinus chalybeus* of Walbaum, (see his edition of Artedi, Vol. II. page 24,) which I suppose is the same with the *Cyprinus bulatmai* of La Cepede, (*Hist. des Poissons*, Tome V. page 490,) and is a fish of the Caspian Sea, which may readily, however, be distinguished from our Indian kind by the great comparative length of the upper jaw, and by the redness of the anal fin.

The *Sarana* sometimes grows to the length of two feet, and is pretty common in the ponds and rivers of India. It is a beautiful fish, but not much valued for eating. The natives of Bengal class it with the following division, calling it *Sarana pungti*, and, in shape and proportion of scales, it has no doubt a considerable resemblance; but then it exceeds the fishes of that division vastly in size, and wants those bright well defined marks of colour by which the others are distinguished. It is of an oblong, but rather deep *form*, compressed, but rather thick make, and rather more protuberant on the back than below. The *colour* is green and silver, the middle of the scales being green, and the edges silver; but towards the back the green greatly prevails, while the scales there are dotted on the middle; and towards the belly the silver is entirely predominant. The head has a purple gloss. The fins are without spots. The eyes are silver-coloured.

The *head* is blunt, oval, small, and smooth. From each angle of the mouth are two minute *tendrils.* The *mouth* extends from the extremity of the nose, descending a little backward, and is small. The *jaws* protrude in opening, the upper one being rather the longest. The upper *lip* has a small bone at each side. The *palate* has several longitudinal wrinkles, peculiar, so far as I know, to this species. The *nostrils* are nearer the eyes than the jaw. The *eyes* are of a good size, and have circular pupils. Each *gill-cover* consists of three plates, rounded behind, and conceals in part its *membrane.*

The *back* slopes, with a slight convexity, from the commencement of the fin to the end of the nose, and the ridge there is rather sharp. The hinder part slopes with a slight concavity. The *belly* forms an arch, which slopes suddenly at the fin behind the vent. The *lateral line* bends, with a slight curvature, from the shoulder. Each *scale* has several fine lines, radiating as it were from the centre.

The *dorsal* fin occupies nearly the middle of the back, and slopes behind: the first of its rays is a thick, short, smooth bone, closely joined to the second, which is a long strong bone, indented along the hinder edge; the others are branched, and the two last are near each other, so that the fin might be said to have only nine rays, with the last divided to the root. The *pectoral* fins are very little shorter than the head, and each has sixteen rays, of which the first is undivided. The *ventral* fins are as large as the pectorals; each has nine rays, of which the first is undivided. The *anal* fin occupies the middle of the tail, and slopes backward: the three first of its rays are undivided, and closely united, the second being longer than the first and shorter than third; the others are branched, the last being divided to the root. The fin of the *tail* is divided into two bluntish lobes, and contains nineteen distinct rays, besides some short compacted ones.

V. Division.—PUNTIUS.

Fishes of the genus *Cyprinus*, without any resemblance to other genera, of a small size, and marked by a few large deep coloured spots.

The fishes of this division among the natives of Bengal have the generic name of *Pungti*, and are very common throughout the Gangetic provinces. They rarely reach four inches in length, and, although all eaten, are considered as of little value. Like those of the last division, they all have oblong bodies, with a moderate depth; and, although all compressed, are of a good width and strong *form*. The *head* is oval and smooth. The *tendrils*, when any exist, are much shorter than the head. The *jaws* protrude in opening. The *lateral line* runs either on or above the middle of the side. The *dorsal* fin is near the middle of the fish, the *anal* is very short, and that of the *tail* consists of two lobes.

45th Species.—CYPRINUS SOPHORE. Plate XIX. Fig. 86.

An opaque *Cyprinus* of the *Puntius* kind, with a diffuse golden spot on the gill-covers; with a black spot at the end of the tail, and another at the root of the dorsal fin; and with the second ray of that fin smooth behind.

The *Sophore*, of the Sangskrit language, is a beautiful little fish, very common in ponds. It is rather deep in proportion to its length, and a little more prominent above than below. Its *colour* above is greenish, with a gloss of silver, and below entirely of the latter colour. Besides the five spots mentioned in the specific character, it has little else remarkable in its colours. The fins have more or less of a yellow tinge, sometimes below, inclining to red. The eyes are stained with red, and of a silver colour.

The *head* is narrower than the body, and small. There are four *tendrils*, so very minute, as often to be scarcely perceptible. The *mouth* is small, and descends obliquely from the extremity of the nose. The upper *jaw* is rather the longest. The *lips* are small, nor has the upper any lateral bones. The *tongue* is sharp, moveable, and smooth, as is the *palate*. The *nostrils* are near the eyes, which are high, and rather large. The *gill-covers* are rounded, and expose their *membranes*.

The *back* slopes up to the commencement of the fin with an arch, and is rounded on the edge; behind the commencement of the fin it slopes down with a slight curve. The *belly* forms a pretty regular curve from the mouth to the end of the anal fin. The *scales* are large, angular behind, dotted near the edge, and marked with lines diverging like the rays of a star.

The *dorsal* fin slopes behind, and contains ten rays, of which the two first are closely united and undivided, the second being twice as long as the first; the others are branched, and the last divided to the bottom. The *pectoral* fins are low, sharp above, and nearly equal in length to the head, each containing fourteen rays, of which the first is undivided, and the others bifid. The *ventral* fins are smaller than the pectorals, and near each other, each having nine rays, of which the first is undivided, and the others branched. The edge of the *anal* fin is slightly arched: it contains seven rays, of which the two first are undivided and united, the second being much the longest; the others are branched, and the last divided to the root. The lobes of the *tail* fin are bluntish: it has nineteen distinct rays, besides some close compacted short ones.

46th Species.—Cyprinus chola.

An opaque *Cyprinus* of the *Puntius* kind, with a diffuse spot of a golden colour on the gill-covers; with one small black spot near the end of the tail; and with the second ray of the dorsal fin smooth behind.

The *Chola* I found in the north-eastern parts of Bengal, where it is found in ponds and other stagnant waters. It is rather deeper in *form* than the *Sophore*, and rather more prominent above than below. The *colour* above is green, with a gloss of silver, below entirely silver. The most remarkable spots are mentioned in the specific character, but the dorsal fin is variegated with dots collected in the form of various spots. The fins are olive, except the ventrals, which are tinged with orange. The eyes are silvery.

The *head* is small, and narrower than the body. At each corner of the mouth is a *tendril.* The *mouth* is small, and descends obliquely. The under *jaw* is the shorter, and along its middle has a slight ridge, on which account, perhaps, the fish should have been placed in the third division, to the fishes in which, however, it has little resemblance. The *lips* are rather fleshy. The *gill-covers* are rounded behind, and edged with a broad membrane.

Both *back* and *belly* form arches. The *lateral line* is bent with a slight curve from the shoulder. The *scales* are large; those on the back are dotted towards the root. There are no scale-like *appendants* above either the ventral or pectoral fins.

The *dorsal* fin slopes backward, and has ten rays, of which the two first are undivided, and very thick, the first being very short: the others are branched. The *pectoral* fins are low, and shorter than the head, and contain about thirteen rays. The *ventral* fins are as long as the pectorals, and each has nine rays, of which the first and last are undivided. The fin behind the *vent* has seven rays, of which the first is short and closely joined to the second, which is undivided; the others are branched, and the last of them divided to the root. The fin of the *tail* contains nineteen distinct rays, besides those which are short and compacted.

47th Species.—CYPRINUS TERIO.

An opaque *Cyprinus* of the *Puntius* kind, with a diffuse golden spot on each gill-cover, and a black one on each side of the tail near its middle; and with three undivided and smooth rays in the dorsal fin.

I have found the *Terio (Teri pungti)* in the north-east of Bengal. Its *form* is less deep than that of the two preceding species, but its back also is more prominent than its belly. The *colour* is silver, with the fore part of the scales on the back greenish. In old individuals the spots on the tail become surrounded with a faint ring. The dorsal fin is yellowish, and has many small spots composed of dark dots. The eyes are silver, stained above with black.

The *head* is of moderate size, and devoid of tendrils. The *mouth* descends obliquely from the extremity of the nose. The *jaws* have scarcely any *lips*. The *eyes* are high and large.

From the fin the *back* slopes forward with a curve, and its edge is convex. The *lateral line* descends straight from the shoulder. The *vent* is behind the middle. The *scales* are large.

The *dorsal* fin has eleven rays, of which the three first are undivided, and closely united, the second being longer than the first, and shorter than the third; the others are branched. The *pectoral* fins are sharp above, low, and shorter than the head, each containing about thirteen rays. Each of the *ventral* fins has nine rays. The fin behind the *vent* has eight rays, of which the three first are undivided, and closely united, the second being longer than the first, and shorter than the third; the others are branched, and the last of them divided to the root. The fin of the *tail* has nineteen distinct rays, besides short compacted ones.

48th Species.—Cyprinus ticto. Plate VIII. Fig. 87.

An opaque *Cyprinus* of the *Puntius* kind, with one black spot on the lateral line above each pectoral fin, and another near the end of the tail; and with the back fin spotted, and its second ray indented behind.

The *Ticto (Tikto sophore)* resembles strongly in *form* the *Sophore*, but wants the golden mark on the gill-covers, and seldom is found above two inches long. I found it in the south-east parts of Bengal. The *colour* above is green, with a gloss of silver, and the latter colour occupies entirely the belly. The fins are pale-green, and, in old individuals, the dorsal, anal, and ventral fins are slightly stained with red; the dorsal is marked with two rows of dark spots. The eyes are silver, with a faint yellow mark before and behind. The spots on the sides are large, well defined, and deep black.

The *head* is blunt, short, rather narrower than the body, and devoid of *tendrils*. The *mouth* is small, and rather low, and descends obliquely. The *jaws* are almost devoid of *lips*, and the under is rather the shorter. There is scarcely any *tongue*. The *palate* is smooth. The *nostrils* are in the middle between the eyes and jaw. The *eyes* are high up, convex, and large. The *gill-covers* are rounded behind, each consisting of two plates, and concealing half of its *membrane*.

The *back*, from the commencement of the fin to the end of the nose, forms a slight arch, with a rounded edge: from the commencement of the fin to the end of the tail is a concavity.

The *throat* has rather a sharp edge; the *belly* below is blunt, and is not so prominent as the back, although it forms a regular arch. The *lateral line* is scarcely distinguishable. The *scales* are large, and marked with lines diverging from a centre; those of the upper parts are dotted towards the roots.

The *dorsal* fin commences at the middle of the fish, slopes backward, and contains ten rays, of which the first is short, undivided, and closely joined to the second, also undivided, but indented behind: the others are branched. The *pectoral* fins are shorter than the head, low, and sharp above, each having twelve rays. The *ventral* fins are still smaller than the pectorals, are placed close to each other, and have each nine rays, of which the first and two last are undivided. The fin behind the *vent* occupies the middle of the tail, slopes backward, and has seven rays, of which the two first are closely united and undivided, the first being a half shorter than the second, and the others are branched. The *tail* fin has twenty distinct rays, besides short compacted ones.

49th Species.—Cyprinus titius.

An opaque *Cyprinus* of the *Puntius* kind, with two black spots on each side, near the lateral line; with no spots on the dorsal fin; and with the second of its rays smooth behind.

The *Titius* (*Tit pungti*) I found in ponds near Calcutta, and, by the natives, it was considered as the same fish with the *Ticto*, called also *Tit pungti;* but, although both kinds have the utmost resemblance, the marks mentioned in the specific character will serve abundantly to distinguish them as species. From the great resemblance, however, I did not think it necessary to take a drawing or particular description of the Calcutta kind.

In the north-east parts of Bengal I saw another fish called

by the same names, and procured a drawing, now in possession of the Bengal Government. It differed in a few particulars from the *Ticto*, but, the drawings being sufficient to point out the differences, I took no notes, and, therefore, until I recover the drawings, I cannot give this fish a specific character, although I call it *Tictis.*

50th Species.—CYPRINUS JOALIUS.

An opaque *Cyprinus* of the *Puntius* kind, with one black crescent-shaped spot on each side of the tail near the end; with the second ray of the dorsal fin smooth; and with the edge of the under lip indented.

I have found the *Joalius (Jaoyali)* in the ponds and ditches of the north-east parts of Bengal. It grows about three inches in length, and is of a brownish-green *colour* above, and silver below. The back fin is dotted. The eyes are golden.

The *head* is large, but not thicker than the body, and very blunt, and the crown slopes downwards towards the nose. Above each corner of the mouth are two tendrils. The *mouth* is low, and extends right backward. The *lips* are fleshy. The *nostrils* are very high. The *eyes* are a little lower, and of a moderate size. There is a vertical dotted line on each *gill-cover*, and this does not conceal its *membrane.*

The edge of the *back* before the fin forms a sharp ridge; that of the *belly* is rounded. The *lateral line* descends from the shoulder, with a slight curve. The *vent* is behind the middle. The *scales* are dotted, and of a moderate size.

The dorsal fin slopes toward the tail, and contains twelve rays, of which the first is short, and closely united to the second, which, like it, is undivided; the others are branched. The *pectoral* fins are shorter than the head, low, and sharp above. The *ventral* fins are still smaller than the pectorals, scarcely reach to the vent, and each has nine rays. The fin behind the *vent* slopes toward the tail, and has eight rays, of which three are undivided, and closely conjoined, the second being longer than the first, and shorter than the third: the others are branched.

51st Species.—Cyprinus pausio.

An opaque *Cyprinus* of the *Puntius* kind, with one black spot near the end of the tail on each side, and with the second ray of the dorsal fin, and the edge of the under lip, smooth.

The *Pausio* (*Paungsi*) I found in the river Kosi; and it has such a strong resemblance to the *Joalius*, that I noted down only the differences, which, besides those mentioned in the specific character, are as follows:

The *head* is narrower. The *dorsal* fin has thirteen rays, of which the last is divided to the root. The fin behind the *vent* has seven rays, the four last being branched, and the last divided to the root.

52d Species.—Cyprinus conchonius.

An opaque *Cyprinus* of the *Puntius* kind, with one black spot near the middle of the tail, and with no spots on the dorsal fin, the second ray of which is indented behind.

I found the *Conchonius* (*Kongchon pungti*) in the ponds of the north-east of Bengal, and in the rivers Kosi and Ami, which are far to the west. It probably, therefore, is found in all the northern borders of the Gangetic provinces.

This is deeper in the *form* than either the *Terio* or *Ticto*, to both of which it has a strong affinity. In Bengal it has clear coloured fins, without spots or dots, and is greenish above, and silver *coloured* below; but the specimens which I saw in Behar had blackish fins; and those which I observed farther west, in the Ami, a stagnant river full of weeds, had the belly stained black, while the fins of the back and tail were yellow, with blackish ends; yet I am persuaded that the fishes which I observed in the three provinces belonged to the same species.

It has no *tendrils*. On each *gill-cover* is a bent line of dots.

The *dorsal* fin contains ten rays, of which the first is short and slender, and closely united to the second, which is strong, undivided, and indented behind; the others are branched. The *pectoral* fins are much smaller than the head, and each contains about ten rays. Each *ventral* fin contains nine rays, of which the first and last are undivided. The fin behind the *vent* has eight rays, of which the three first are undivided, and closely united, the second being longer than the first, and shorter than the third: the others are branched. The fin of the *tail* contains nineteen distinct rays, besides short compacted ones.

53d Species.—CYPRINUS PUNTIO.

An almost opaque *Cyprinus* of the *Puntius* kind, with a black ring round the end of the tail, and a black spot at the roots of the dorsal and anal fins.

This fish I found in the ponds and ditches of the southern parts of Bengal, where it is called *Pungti;* and it was brought to me as the young of the 30th species, or *Catla;* but although it resembles the *Catla* in form, and although several young fishes differ very much in colour from the adults, yet the differences in the number of rays, which support the fins of the two animals, renders it highly improbable that they should belong to the same species.

The *form* of the *Puntio* is very similar to that of the *Catla*. The back and tail are of a greenish-brown *colour*, almost diaphanous. The abdomen and gill-covers are silvery white. On each side is a diffuse stripe of greenish gold colour. The pectoral, ventral, and tail fins, are glassy; the dorsal and anal are deep yellow. The eyes are silver.

The *lateral line* is indistinct. The *scales* are large.

The *dorsal* fin has nine rays, of which the first is undivided. Each *pectoral* fin has twelve rays, each *ventral* eight, and the *anal* six. The lobes of the *tail* fin are equal and sharp.

54th Species.—CYPRINUS PHUTUNIO.

A somewhat diaphanous *Cyprinus* of the *Puntius* kind, with five black spots on each side; and with the second ray of the back fin indented behind.

The *Phutunio (Phutuni pungthi)* I found in the ponds of the north-east of Bengal. It strongly resembles the next described species, the *Gelius*, so that I have not taken a very particular description.

The *Phutunio* is deep in the *make*, and except the gill-covers and membrane covering the viscera, which are silver *coloured*, the whole animal is slightly diaphanous. On each side are two long black spots placed with their longest axis vertical; and, besides these, there are three other spots, which appear as if deeply seated, and as if they were seen through the skin. The fin of the back is spotted, and the ventrals are red.

There are no *tendrils*. The *mouth* is very small.

The *back*, from the fin to the head, slopes more rapidly than any other of the *Puntius* kind. The *lateral lines* are very faint. The *scales* are large.

The *dorsal* fin contains ten rays, of which the second is undivided and indented behind. Each *pectoral* fin seems to have only nine rays, but of this number I am not certain. Each *ventral* fin has only eight rays. That behind the *vent* has seven.

55th Species.—Cyprinus gelius.

A yellowish diaphanous *Cyprinus* of the *Puntius* kind, with the gill-covers, the membrane covering the viscera, and a longitudinal stripe along the spine, all of a silver colour; with several large spots on each side; and with the second ray of the back fin indented behind.

The *Gelius (Geli pungti)* I found in the ponds and ditches of the north-east parts of Bengal, where it grows to about two inches in length. It is deep in the *make,* and protuberant on the back. On the sides are scattered several large spots. The back fin is yellowish, with a black spot at its root, as has also each ventral: the anal is spotted at the lower part.

The *head* is of a moderate size, and without *tendrils.* The *mouth* descends obliquely from the extremity of the head, and is small. There are no *lips.* The *eyes* are large and high.

The *back* slopes from the fin with a rounded edge. The *lateral line* runs straight above the middle of the side. The *scales* are large and dotted.

The *back* fin has ten rays, of which the first is short and closely united to the second, which, like it, is undivided; the others are branched. The *pectoral* fins are low, shorter than the head, and sharp above, with rays so slender that I cannot ascertain their number. Each *ventral* fin has nine rays: the *anal* has eight, of which the three first are undivided and closely united, the second being longer than the first, and shorter than the third: the others are branched.

56th Species.—Cyprinus canius.

A reddish diaphanous *Cyprinus* of the *Puntius* kind, with the gill-covers, and the membranes covering the viscera and spine, like silver; with several large black spots on each side; and with the second ray of the back fin indented behind.

This is a beautiful fish, about an inch and a half in length, which I have found in the ponds and ditches of the north-east parts of Bengal. It is of rather a deep *form*, and more protuberant above than below.

The *head* is rather large, oval, and without *tendrils*. The *mouth* descends obliquely from the extremity of the head, and is small. There are no *lips*. The *eyes* are large and high.

The slope of the *back*, from the fin to the head, is rounded on the edge. The *lateral line* is straight. The *scales*, in proportion, are very large.

The *back* fin slopes behind, and contains ten rays, of which the first is short, and closely united to the second, which is simple; the others are branched. The *pectoral* fins are low, and shorter than the head, each having about eight rays. Each of the *ventral* fins has nine rays; and the *anal*, which slopes behind, has seven, of which the first two are undivided and closely united, the first being short: the others are branched.

VI. Division.—Cyprinus danio.

Fishes of the genus *Cyprinus*, without any resemblance to other genera, of a small size, and marked with longitudinal lines, without remarkable spots.

Several species of this division, especially such as have the strongest resemblance to the *Minnow, (Cyprinus phoxinus,)* are, by the Bengalese, called *Dhani*, or some variation of that name, with different additions; and it is on this account that the name Danio has been adopted. In an economical view, the fishes of this division are of still less value than the preceding, as they are equally insipid and smaller; nor does their number any where compensate for these defects.

57th Species.—Cyprinus dangila.

A *Cyprinus* of the *Danio* kind, with blue stripes on the sides, uniting together into a net-work; with the body much compressed; and with four tendrils, of which two are rather longer than the head.

I found the *Dangila* among the rocks and stones which compose the bottom of several clear mountain streams, south from Mungger. It is a beautiful fish, about the length of the finger, very much compressed, and of a long *form*, more prominent below than above. The back is of an olive *colour*, and the belly silver, having on each side several blue lines, that form a beautiful net-work. Behind the gill-cover, and above each pectoral fin, is a large blue spot. The fin behind the vent is edged with green, and has two or three stripes of the same colour. The eyes are silver, with some green on their upper sides.

The *head* is half oval and small. Two of the *tendrils*, proceeding from before the nostrils, are a little shorter than the head; and the two from the corners of the mouth are a little longer. The *mouth* descends obliquely, and is of moderate size. The *jaws* do not protrude in opening, and the under one is the longest. The *eyes* are on the sides of the head, and of moderate size, with the longest axis of their pupil vertical. The *gill-covers* are dotted.

The *back*, from the head to the commencement of the fin, scarcely rises. The *belly* is prominent. The *lateral line* is bent down parallel to the belly. The *vent* is behind the middle. The *scales* are rather large in proportion. Above each ventral fin is a scale-like *appendant*.

The *dorsal* fin, placed far back, and ending sharp behind, has thirteen rays, of which the two first are closely united and undivided, and the others are branched, the last being divided to the root. The *pectoral* fins are low, sharp above, and nearly of the same length with the head, and each contains about twelve rays. The *ventral* fins are smaller than the pectoral, do not extend to the vent, and each has only seven rays, of which the last is divided to the root. The fin behind the vent has seventeen rays, of which the three first are undivided, the second being longer than the first, and shorter than the third. The tail fin ends in a small sinus, and contains twenty distinct rays, besides four short compacted ones on each side, in all twenty-eight.

58th Species.—CYPRINUS RERIO.

A *Cyprinus* of the *Danio* kind, with several blue and silver stripes on each side; with the body much compressed, and with four tendrils, of which two are a little longer than the head.

This beautiful fish I found in the Kosi river, where it grows to about two inches in length. The *form* is oblong, projecting more below than above. The *colour* above inclines to olive, and below to silver, with several longitudinal blue and silver stripes alternating on each side. The fin behind the vent is barred, and that of the tail striped with blue. The eyes are silver.

The *head* is small and half oval. The *mouth* descends obliquely backward, and is of moderate size. The *jaws* protrude in opening, the under one being rather the longest. The two *tendrils* before the nostrils are very short; the two from the corners of the mouth are a little longer than the head. The *eyes*, placed on the sides of the head, are of moderate size, with circular pupils. The *gill-covers* do not conceal their *membranes*.

The *back*, from the fin to the head, is straight, and the *breast* convex. The *lateral line* is scarcely observable. The *scales* are of moderate size. The *vent* is rather behind the middle. There are no scale-like *appendants* above the lower fins.

The fin of the *back* is behind the middle, and is rounded. It has eight rays, of which the first is undivided. The *pectoral* fins are low, and sharp above: each contains about nine rays; but they are indistinctly visible. The *ventral* fins are small, each containing only seven rays. In the fin behind the *vent* there are sixteen. The *tail* fin ends in a crescent, or sinus; but its rays, on account of their slenderness, and the brightness of the colours, cannot be counted.

59th Species.—Cyprinus chapalio.

A *Cyprinus* of the *Danio* kind, with three blue stripes on each side, with the body slightly compressed, and with two tendrils shorter than the head.

The *Chapalio* is a very beautiful fish, little longer than the *Rerio*, to which it has a strong resemblance. It is found in the ponds of Bengal. The *form*, like that of the *Minnow*, (*Cyprinus phoxinus*,) is oblong, very slightly compressed. The blue *coloured* stripes on the sides are separated by intermediate ones of gold or silver. The back is dusky, the belly white. There are five blue stripes on the fin of the tail, and three on that behind the vent.

The *head* is small, and blunt. The *tendrils*, which proceed from the corners of the mouth, are slender, and shorter than the head. The *mouth* descends obliquely backward. The under *jaw* is rather the longest. The *gill-covers* end in a point.

The *back* forms a slight arch. The *belly* projects considerably. The *lateral line* is straight, high up on the side, and smooth. The *vent* is near the middle. The *scales* are large, and adhere firmly.

The *back* fin is behind the middle, its edge is arched, and it contains nine rays, of which the first is short and undivided, and the others branched. The *pectoral* fins are low, shorter than the head, and sharp above; each has twelve rays. The *ventral* fins are near each other, and very small, each having seven rays. The *anal* fin is near the vent, and nearly of equal width throughout, and has thirteen rays, of which the two first are undivided, and the others branched. The fin of the *tail* has a slight notch in the middle, and contains eighteen distinct rays, besides some short compacted ones at each side.

60th Species.—CYPRINUS DANRICA. Plate XVI. Fig. 88.

A *Cyprinus* of the *Danio* kind, with one dark dotted stripe below the lateral line; with the form slightly compressed; with four tendrils, of which two are very long; and with the first ray of each ventral fin lengthened into a bristle.

The *Danrica* is found in the ponds and ditches of Bengal, and strongly resembles the *Minnow (Cyprinus phoxinus)* in size and appearance. The *form* is oblong, very slightly compressed. The *colour* above is green, and dotted, below silver. There is only one longitudinal stripe on each side. The lower fins are yellowish. The eyes are silver, dotted with black.

The *head* is flat above, narrower than the body, and short. From each angle of the mouth are two slender *tendrils*, of which the upper is minute, and the under extends to the ventral fins. The *mouth* commences high, descends much backward, and is small. The *jaws* have scarcely any *lips*, and protrude much in opening. The *nostrils* are near the eyes, and circular. The *eyes* are high up, and far forward, and are small. The *gill-covers* are rounded, and half conceal their *membranes*.

The *back* is convex on the edge, and rises little towards the fin. The *belly* is rather sharp-edged. The *lateral line* is straight, higher than the middle, and in a golden stripe. The *vent* is behind the middle. The *scales* are large in proportion, and easily removed.

The *back* fin is far behind the middle, and rounded. It contains eight rays, of which the first is short, and, together with the second, is undivided, while the others are branched. The *pectoral* fins are as long as the head, low, and sharp above; each contains thirteen rays, of which the first is undivided, and the others branched. The *ventral* fins are near each other, and as long as the pectorals; each has eight rays, of which the first is by much the longest and undivided, ending in a bristly point, projecting beyond the membrane; the last also is undivided, and the intermediate ones branched. The *anal* fin is near the vent, and triangular, and has eight rays, of which the first is short, and, with the second, is undivided, the others being branched. The fin of the *tail* is large, is divided into two equal lobes, and contains nineteen distinct rays, besides short compacted ones on each side.

61st Species.—Cyprinus jogia.

A *Cyprinus* of the *Danio* kind, with one dark dotted stripe below the lateral line; with the form much compressed; with four tendrils, of which two are very long; and with the first ray of each ventral fin no longer than the membrane.

The *Jogia (Jongja)* I found in the Kosi along with the *Rerio*, to which it has a considerable resemblance, and by the fishermen is considered as belonging to the same genus: but, on the whole, its affinity with the *Danrica* is stronger; and, besides the circumstances mentioned in the specific character, it differs from the description of the *Danrica* in only the following particulars:

The *eyes* are entirely silver-coloured. The *belly* is more arched. The *pectoral* fins are longer than the head, and each contains about ten rays. The rays in each *ventral* and in the anal fin are seven in number.

62d Species.—CYPRINUS SUTIHA.

A *Cyprinus* of the *Danio* kind, with one silver stripe along the lateral line; with the form much compressed; with four tendrils, of which two are very long; and with the first ray of each pectoral and ventral fin produced in a slender point beyond the membrane.

The *Sutiha* is found in the ponds of the Gorakhpur district, and has the utmost affinity to the two last described species, from which, however, it may at once be distinguished by wanting the dotted stripe on its sides. Besides this and the circumstances mentioned in the specific character, it differs from the *Jogia* in the following particulars:

The lower part of the *body* is diaphanous, with a silver membrane lining the visceral cavity. The under fins are pellucid.

63d Species.—CYPRINUS DANICONIUS. Plate XV. Fig. 89.

A *Cyprinus* of the *Danio* kind, with one dark dotted stripe below the lateral line; with the form slightly compressed; with no tendrils; with nine rays in the back fin, and seven in that behind the vent; and with the lateral line straight.

The *Daniconius (Danikoni)* is pretty common in the rivers of Southern Bengal. It is a beautiful fish, three or four inches long, and in *colour* has a strong resemblance to the three last described. The *form* is long, and tapers equally to both ends. On each side are two adjacent longitudinal stripes, the upper gold *colour*, the other dark from numerous dots; but in many individuals these are not so numerous as in the specimen from which the drawing was taken. The fins are yellowish without marks.

The *head* is oval, blunt, and of moderate size. The *mouth* is small, and descends a very little backward. The *jaws* have no *lips*, and protrude in opening, the under one being rather the longest, and ending in a sharpish point, while the upper ends in a notch. The *eyes* are high, convex, and rather large, with circular pupils. The *gill-covers* are rounded, and consist of two bony plates, which do not entirely conceal the *membranes*.

The *back*, from the fin to the nose, forms a regular slight curve. The *lateral line* is straight and above the middle. The *vent* is behind the middle. The *scales* adhere slightly, are large, and are dotted and marked with lines diverging from the centre.

The fin of the *back* is rather behind the middle, slopes back-ward, and contains nine rays, of which the first is short, and with the second is undivided, the others being branched. The *pectoral* fins are low, shorter than the head, and sharp above, each having thirteen rays. The *ventral* fins are small, and each has nine rays, of which the first is undivided. The *anal* fin is under the middle of the tail, slopes backward, and contains seven rays, of which the first is short, and, like the second, is undivided, the remainder being branched, and the last divided to the root. The fin of the *tail* is divided into two equal lobes, and contains nineteen distinct rays, besides several short compacted ones.

64th Species.—Cyprinus anjana.

A *Cyprinus* of the *Danio* kind, with two dotted longitudinal stripes on each side; with the form somewhat compressed; with no tendrils; and with nine rays in the dorsal fin, and seven in the anal.

The *Anjana (Angjana)* I found in the ponds of the Puraniya district. It is a small fish, very like the *Daniconius*, and is of an oblong *form*, and thick make. The *colour* of the upper parts is green, and of the lower silver. The upper dotted stripe runs straight along the middle of the side, and is edged above with gold: the under is parallel to the belly, and is interrupted. The fins are yellow, with dark dots. The eyes are silver

The *head* is narrower and more compressed than the body, and it is flat above, and blunt. The mouth is of a moderate size, and descends backward. The *jaws* are flattened, without lips, and protrude in opening, the under one being rather the longest, and ending in a sharp point, while the upper one ends in a notch. The *eyes* are high, and of moderate size, and have circular pupils. The *gill-covers* are somewhat angular, and expose part of their *membranes.*

The edge of the *back* is rounded, and slopes suddenly at the nape. The *lateral line* is low, and runs parallel to the *belly.* The *vent* is behind the middle. The *scales* are large, adhere but slightly, and are both dotted and marked with diverging lines. Above each ventral fin is a scale-like *appendant.*

The *dorsal* fin, situated behind the middle, has nine rays, of which the first is short, and, with the second, is undivided, the others being branched, and the last divided to the root. The *pectoral* fins are shorter than the head, and each has about thirteen rays. The *ventral* fins are small, and each has nine rays. The *anal* fin slopes backward, and has seven rays like those of the dorsal. The *tail* fin is divided into two lobes, and contains about nineteen distinct rays, besides some short compacted ones.

65th Species.—Cyprinus rasbora. Plate II. Fig 90.

A *Cyprinus* of the *Danio* kind, with one slightly dotted longitudinal stripe on each side; without tendrils; with the form slightly compressed; with nine rays in the fin of the back, and eight in that behind the vent; and with the lateral line bent parallel to the belly.

The *Rasbora* is found in the ponds of Bengal, and has a very strong resemblance to the *Daniconius*, but is rather larger. The *form* resembles that of the head of a lance, but is only slightly compressed. The *colours* above are green, with many small dots, and below silver. On each side are two longitudinal faint contiguous stripes, the upper golden, the under dotted with black. The fin of the tail is yellow, with the hinder edge stained black.

The *head* is oval, sharpish, and of moderate size. The *mouth* is small, and descends a very little backward. The *jaws* protrude in opening; the under is rather the longest, and ends in a sharp point; the upper ends in a notch. The *eyes* are high, rather large, and convex, and have circular pupils. Each *gill-cover* has three plates, and is rounded and dotted.

The *back* before the fin forms a curve that slopes rather suddenly at the nape. The *lateral line* is low. The *vent* is behind the middle. The *tail* contracts suddenly at the fin behind the vent. The *scales* are large, adhere slightly, and are dotted and marked with lines diverging from a centre.

The *dorsal* fin is near the middle, slopes backward, and has nine rays, of which the first is short, and, like the second, is undivided, the others being branched, and the last divided to the root. The *pectoral* fins are low, shorter than the head, and sharp above: each has thirteen rays. The *ventral* fins are small, and close together, each having nine rays, of which the first and last are undivided. The *anal* fin, placed near the vent, slopes with a concave edge, and contains eight rays, of which the three firs tare undivided, the second being longer than the first, and shorter than the third: the others are branched; and the last is divided to the root: The fin of the *tail* consists of two lobes, of which the under one is the longest; and both contain nineteen distinct rays, besides several short compacted ones.

66th Species.—CYPRINUS ANGRA.

A *Cyprinus* of the *Danio* kind, with the sides stained yellow, and marked with one broad longitudinal dark dotted stripe; with two minute tendrils; and with the form compressed.

The *Angra* is found in the Brahmaputra river, and is about a span in length. Although, from its colours, it belongs to a different division, it has, in every other respect, the utmost resemblance to the *Morala*, which will be next described; and, indeed, differs from the description of that only in the few particulars mentioned in the specific character, and in the following circumstances:

The *eyes* are red. The minute first ray of the *dorsal* fin is wanting. All the fins are dotted with black.

VII. Division.—MORULIUS.

Fishes of the *Cyprinus* genus, with no resemblance to other genera, of a small size, and with the colours of the belly and back indented into each other.

All the fishes of this division have the strongest affinity to each other, yet, among the natives, their names are quite distinct. I have formed the name of the division from one of those given to the most common species. The 20th species, or *Dero*, differs in nothing from this division, except in having the ridge on the under jaw like a *Mugil*.

67th Species.—CYPRINUS MORALA. Plate XVIII. Fig. 91.

A *Cyprinus* of the *Morulius* kind, with four tendrils; with thirteen rays in the fin of the back; and with eight in that of the vent.

The *Morala* I have found in the fresh waters of Bengal, where it reaches to the size of a small herring. The *form* is long and compressed. The upper parts are of a brownish-green *colour*, with many dark dots, the under are silver. The fins are of the same colour with the adjacent parts. The eyes are silver.

The *head* is small, long-oval, and dotted on the nose. The *tendrils* are minute. The *mouth* is under the nose, and small, and opens directly backward. The *jaws* protrude in opening, and are covered by pendulous *lips*, of which the under one is indented on the edge. The *nostrils* are close to the *eyes*, which are high, flat, and of moderate size, with the longest axis of the pupil vertical. The *gill-covers* do not conceal their *membranes.*

The *lateral line* runs straight along the middle of the side. The *vent* is behind the middle. Before the fin the *back* slopes straight down to the nose. The *scales* are large, adhere firmly, and are surrounded by a margin. Above each ventral fin is a scale-like *appendant.*

The *dorsal* fin is before the middle, and its edge forms a concave curve. It contains thirteen rays, of which the second is longer than the first, and shorter than the third, all the three being undivided, but the third has joints: the others are branched, and the last is divided to the root. The *pectoral* fins are shorter than the head, each containing about sixteen rays. The *ventral* fins reach, with their ends, to the vent, and each has nine rays. The edge of the *anal* fin is concave, and it contains eight rays like those of the back. The fin of the *tail* consists of two lobes, of which the upper is the longest and sharpest; and it contains nineteen distinct rays, besides several short compacted ones.

68th Species.—Cyprinus pausius.

A *Cyprinus* of the *Morulius* kind, with four tendrils; with fourteen rays in the fin of the back; and with seven in that of the vent.

The *Pausio (Paungsi)* I found in the Kosi river. It is a smaller fish than the *Morala*, and its colours fainter; but, in other respects, except the number of rays in the back and anal fins, it scarcely differs from that fish.

69th Species.—Cyprinus musiha.

A *Cyprinus* of the *Morulius* kind, without tendrils; with thirteen rays in the fin of the back, and eight in that behind the vent; and with the body compressd.

The *Musiha* I found in the Ganges at Patna. It differs from the description of the *Morala* in nothing but the want of tendrils, and those of the *Morala* are so minute, that I have some doubts of their being a sufficient mark of distinction.

VIII. Division.—Cabdio.

Fishes of the genus *Cyprinus*, having no resemblance to other genera, and of a small size; without spots, stripes, or other remarkable distinction of colour; and with a form very considerably compressed.

The small fishes composing this division are of little importance either from number or quality; nor have the natives any generic appellation common to the greater part. I have adopted one of the many names *(Kavdi)* given to the most common species. None of them, that I have seen, have tendrils.

70th Species.—Cyprinus jaya.

A *Cyprinus* of the *Cabdio* kind, with eight rays in the dorsal fin, about thirteen in each pectoral, eight in each ventral, and nine in that behind the vent; and with two lateral lines on each side.

The *Jaya* is found in the rivers of Northern Behar, and grows to about two inches in length. The *form* is long and compressed. The *colour* is silver, with a greenish back, and no spots either on the body or fins. The eyes are silver.

The *head* is oval, small, and blunt. The *nose* is fleshy, and projects a little beyond the *mouth*, which is small and arched. The *jaws* have scarcely any lips, and protrude in opening, the upper being the longest, and entire at the end. The *nostrils* are near the eyes, and each has only one aperture. The *eyes* are high, near the middle of the head, and of moderate size. The *gill-covers* are rounded.

The *back* has a convex edge, with a pale line before the fin. The *shoulder-bones* are covered with scales. The upper *lateral line* is straight, the under parallel to the belly. The *vent* is behind the middle. The *scales* adhere slightly, and are small. There are no scale-like *appendants* above the lower fins.

The *dorsal* fin is behind the middle: the first and second of its rays are undivided; the first being short: the others are branched, and the last divided to the root. The *pectoral* fins are low, sharp above, and smaller than the head. The *ventral* fins do not extend to the vent, and their rays are undivided. The *anal* fin slopes backward: the first and second of its rays are short, and the last is divided to the root. The *tail* fin is divided into two lobes, of which the upper is the bluntest. It contains seventeen distinct rays, besides several short compacted ones.

71st Species.—Cyprinus mola. Plate XXXVIII. Fig. 92.

A *Cyprinus* of the *Cabdio* kind, with nine rays in the dorsal fin, seven in that behind the vent, fifteen in each pectoral, and nine in each ventral; and with two lateral lines on each side.

This is the most common and most generally diffused species of this division with which I am acquainted, and is found in both ponds and fresh water rivers in every part of the Gangetic provinces. It grows to three or four inches in length.

The *form* resembles the head of a lance, and is much compressed. The *colours* above are green, and below silvery, but the body is somewhat diaphanous, so that membranes like silver, which line the visceral cavity and spine, are visible. The eyes are like silver.

The *head* is blunt, oval, and of moderate size. The *mouth* terminates the head, descends obliquely backward, and is of moderate size. The *jaws* have no lips, and scarcely protrude in opening; the upper is rather the shortest, ends in a notch, and has a membranous veil projecting from its inner side; the under one is rather sharp. The *tongue* is sharpish and smooth, as is the palate. Each *nostril* has only one aperture, and is situated in the middle between the eye and nose. The *eyes* are on the middle of the cheeks, and are of moderate size and globular. The *gill-covers* consist of three bones, of which the uppermost is by much the largest, is rounded behind, and marked by a bent line of dots. The *membrane* of each is exposed to view. The bones supporting the *gills* on their inside have a row of tubercles.

The *back* is arched with a gentle curve, having before a declivity greater than that of the head: its edge is convex. The *tail* slopes suddenly at the anal fin. The upper *lateral line* runs straight above the middle of the side; the under runs downward, terminating above the ventral fins. The *vent* is behind the middle. The *scales* adhere loosely, and are small.

The *dorsal* fin is nearly in the middle of the fish, and slopes backward: the first of its rays is short, and closely united to the second, both being undivided; the others are branched, and the last is divided to the root. The *pectoral* fins are very low, shorter than the head, and sharp above. The *ventral* fins are shorter than the pectorals: the first and last rays of each are undivided, and the intermediate ones branched. The *anal* fin slopes backward, and its rays are like those on the back. The *tail* fin consists of two lobes, the under one being rather the longest, and contains nineteen distinct rays, besides several short compacted ones.

72d Species.—CYPRINUS HOALIUS.

A *Cyprinus* of the *Cabdio* kind, with nine rays in the dorsal fin, ten in the anal, and nine in each ventral; with one lateral line on each side; and with the belly silver-coloured.

The *Hoalius (Hayali)* is found in the Atreyi, and other rivers in the northern parts of Bengal, and grows to about six inches in length. It has some affinity to the *Leusiscus* or *Dace*, and *Dobula*, but is smaller. The description I took was very imperfect.

The *form* is much compressed, straightish above, and prominent below. The *colours* are above green, and below silver.

The *head* is small. The *jaws* are equal in length.

The *lateral line* is bent parallel to the edge of the belly.

The fin of the *tail* is divided into two lobes.

73d Species.—CYPRINUS BORELIO.

A *Cyprinus* of the *Cabdio* kind, with nine rays in the dorsal, eleven in the anal fin, and eight in each ventral; with two lateral lines on each side; and with the belly yellow.

In the Gangetic provinces the Borelio *(Boreli)* is a pretty common fish, and grows to about four inches in length.

The *form* is long and compressed. The *colour* above is silver, with a tinge of green, and below a deep yellow, extending to the lower fins. The eyes are silver.

The *head* is small. The *nose* projects a little beyond the *mouth*, which is very small, and extends directly backward. The upper *lip* is fleshy. The *jaws* protrude in opening, the upper being the longest, and having a notch in the middle. Each *nostril* has only one aperture, situated near the eye. The *eyes* are far forward on the head, and of moderate size.

The edge of the *back* is convex. The *shoulder-bones* are shining, and devoid of scales. The upper *lateral line* is straight; the under parallel to the belly. The *vent* is behind the middle. The *scales* adhere slightly, and are of moderate size. There are no scale-like *appendants* above the lower fins.

The *dorsal* fin is behind the middle, and slopes backward: the first of its rays is short, and united to the second, which is undivided; the others are branched, and the last of them is divided to the root. The *pectoral* fins are low, and sharp above; each has about thirteen rays. The *ventral* fins are small, and do not reach to the vent; the first and last ray of each is undivided. The *anal* fin slopes behind, and its rays are like those on the back. The *tail* fin is divided into two lobes, and contains eighteen distinct rays, besides short compacted ones.

74th Species.—CYPRINUS SOLIO.

A *Cyprinus* of the *Cabdio* kind, with nine rays in the dorsal fin, eleven in the anal, and eight in each ventral; with two lateral lines on each side; and with the belly like silver.

The *Solio (Soli)* is found in the Kosi river, and differs from the *Borelio* only in the *colour* of the belly, which in no part inclines to yellow, and in having scale-like *appendants* above both pectoral and ventral fins. It is rather of a smaller size.

75th Species.—CYPRINUS COSUATIS.

A *Cyprinus* of the *Cabdio* kind, with ten rays in the dorsal and seven in the anal fin; with one lateral line on each side; and with an opaque body.

I found the *Cosuatis (Koswati)* in the Kosi river. It has a great resemblance to the *Terio*, (47th species,) but wants the fine spots by which that fish is distinguished. The *form* is deep, compressed, more prominent on the back than below. The *colour* is silver, with a greenish back, and the scales on that part are dotted towards the root. The ventral fins are red, all the others are yellowish, and those of the back and behind the vent are stained with black. The eyes are silver, stained above with black.

The *head* is oval, and of a moderate size. The *mouth* opens at the extremity of the head. The *jaws* protrude much in opening, and have no lips: the under one is by a little the shorter. The *eyes* are high and large. The *gill-covers* are generally dotted.

Before the fin the *back* slopes downward in a curve, and has a rounded edge. The *lateral line* is straight, and descends along the side from the shoulder. The *vent* is behind the middle. The *scales* are large, and there are no scale-like *appendants* above the lower fins.

The *dorsal* fin is somewhat rounded: the first of its rays is short, and, with the second, which is smooth, is undivided; the others are branched. The *pectoral* fins are low, much shorter than the head, and sharp above. The *ventral* fins are small: each has nine rays. The rays of the anal fin are like those of the back. The fin of the tail is divided into two lobes.

76th Species.—CYPRINUS GUGANIO.

A *Cyprinus* of the *Cabdio* kind, with ten rays in the dorsal, and seven in the anal fin; with scarcely any lateral line, and with a diaphanous body.

The *Guganio (Gugani)* is probably found in most of the rivers and ponds of the Gangetic provinces, as I have seen it in both the Brahmaputra and Yamuna, the extreme rivers of that territory. It scarcely exceeds an inch and a half in length, and has a considerable affinity to the *Mola*, (Species 71.) The *form* is compressed, deep, and more prominent below than above. It is diaphanous, so that eleven ribs on each side may be seen through the scales. The back is dotted: the head, eyes, the membrane lining the visceral cavity, and the spine, are *coloured* like silver. The fins are diaphanous, the fore part of the dorsal being dotted.

The *head* is oval, small, blunt, and dotted. The *mouth* is small. The *jaws* protrude in opening; the upper is the longest. The *nostrils* cannot be clearly distinguished. The *eyes* are far forward on the sides of the head, and large. The *membrane* of each gill-cover contains four rays.

Before the fin the *back* slopes downward. The *scales* are large in proportion, and adhere firmly. The *lateral line*, if any exists, is very faint. The *vent* is behind the middle.

The *dorsal* fin is near the middle: the first of its rays is short, the second strong, undivided, and indented behind: the others are branched. Each of the *pectoral* fins has about twelve rays; but the lower ones are so slender as to be scarcely visible. Each of the *ventral* fins has nine rays, of which the first and last are undivided. The first ray of the *anal* fin is short, and the last divided to the root. The *tail* fin is divided into two lobes.

77th Species.—CYPRINUS COTIO. Plate XXXIX. Fig. 93.

A *Cyprinus* of the *Cabdio* kind, with ten rays in the dorsal fin, thirty-six in the anal, and twelve in each ventral.

The *Cotio (Koti)* in the ponds and ditches of Bengal is a pretty common fish, grows to about four inches in length, and is full of bones.

The *form* is deep, very much compressed, and somewhat like a trapezium. The *colour* is above green, and below white with a silver gloss, and somewhat diaphanous. At the commencement of the dorsal fin is a small black spot, and under the fore part of the lateral line there is a row of five or six others, around which is a bluish shining depression. The eyes are silver, stained above with dark green.

The *head* is oval, blunt, and of moderate size: above it is covered by a naked bone, on which is a depression in form of a parallelogram. The *mouth* terminates the nose, descends obliquely, and is small. The *jaws* are nearly equal, have no *lips*, and protrude in opening. The *nostrils* are near the eyes, which are large, and high up, with circular pupils. The *membranes* of the gill-covers are half exposed. The bones supporting the *gills* have on the inside a row of tubercles.

Before the fin the *back* forms a sharp ridge, and slopes more than the head. The *lateral line* runs straight above the middle of the side. The *vent* is before the middle. The *scales* adhere slightly, are small, and on the back are dotted.

The dorsal fin is near the middle, and slopes backward: the first of its rays is short; the second, like the first, is undivided, and is indented behind; and the others are branched, the last being divided to the root. The *pectoral* fins are low, nearly as long as the head, and sharp above; each contains sixteen rays, of which the first is undivided. The *ventral* fins are small, and the first ray of each is very short, and adheres to the second, which is undivided. The *anal* fin occupies almost the whole length of the tail, and slopes backward: the three first of its rays are undivided, the second being longer than the first, and shorter than the third; the others are branched, and the last is divided to the root. The *tail* fin is divided into two sharp lobes, and contains nineteen distinct rays, besides several compacted ones.

78th Species.—CYPRINUS DEVARIO. Plate VI. Fig. 94.

A *Cyprinus* of the *Cabdio* kind, with eighteen rays in the fin of the back, and an equal number in that behind the vent.

The *Devario* (*Debari*) is also a pretty common fish in the rivers and ponds of Bengal: in the former it is sometimes a span in length; but in the latter, where it is most common, it seldom reaches half that size. It is a very beautiful animal, of a green *colour* above, with generally many large irregular golden spots, and a golden gloss: below it is like silver; and on each side of the tail, extending to the end of the fin, is a bluish stripe, changing in certain lights to green. The fins are yellow; the eyes are white, with a golden ring round the pupil. The *form* is much compressed, and deep, with the belly prominent.

The *head* is oval and blunt, and descends with less slope than that of the back. The *mouth* is at the extremity of the head, descends obliquely, and is large. The *jaws* have no *lips*, protrude in opening, and are nearly equal in length: in the under is a slight roughness, which may perhaps be considered as teeth; but it is scarcely distinguishable, except to the touch. The *tongue* is sharpish, tied down, and smooth. The *palate* is smooth, with a projection on each side. Each *nostril*, in the middle, between the point of the nose and eye, has only one aperture. The *eyes* are rather large, high, and globular. The *gill-covers* are rounded, and expose part of their *membranes*. The bones supporting the *gills* are smooth on the inside.

The *back* has a convex ridge, and forms a regular curve from the nape to the end of the tail. The *ridge below* also is convex, and the curve very prominent before the *vent*, which is behind the middle. On each side there is one *lateral line* parallel to the belly. The *scales* are large and dotted, adhere firmly, and are marked with concentric lines.

The *dorsal* fin is nearly in the middle, and slopes very little: the first two of its rays are undivided, and the first of them is very short: the others are branched. The *pectoral* fins are low, nearly as long as the head, and sharp above: each has ten rays. The *ventral* fins are small: each contains eight rays. The *anal* fin occupies the middle of the tail, and slopes a little backward: its rays are like those on the back. The *tail* fin is divided into two lobes, and contains nineteen distinct rays, besides several short compacted ones.

79th Species.—Cyprinus dancena.

A *Cyprinus* of the *Cabdio* kind, with six rays in the dorsal, and twenty in the anal fin.

The *Dancena (Dhankena)* is a fish about an inch and a half long, which I found in the estuary below Calcutta, near which city both it and the third species *(Cyprinus Laubuca)* are called by the same name, having a strong resemblance. The *form* is very much compressed, with the back rising gently to the commencement of the fin, and the lower edge very prominent. The *colours* are green above, and silver below.

The *head* is half oval, flat above, broader than the body, and small. The *mouth* is small, and descends backward obliquely. The upper *jaw* is the longest. The *lips* are fleshy. The *eyes* are high, and far forward.

The *lateral line* runs straight above the middle of the side. The *vent* is near the middle. The *scales* are large in proportion, but so thin, as scarcely to be visible without a glass, except on the belly.

The *dorsal* fin is far back. The *pectoral* fins are of moderate size, are placed high on the side, and are sharp above: each has ten rays. The *ventral* fins are small, each having six rays. The *tail* fin is rounded, and has sixteen rays.

IX. Division.—CYPRINUS GARRA.

Fishes of the *Cyprinus* kind, of a small size, with no affinity to another genus; without remarkable spots or stripes; and with a long form, very little, if at all, compressed.

The little fishes of this division live among stones in rapid mountain streams, or in rivers with pure sandy bottoms, and, like many of the *Loaches*, *(Cobitis,)* usually lie close to the bottom, so that several of them are by the natives called *Balitora*, or sand-diggers; and the name *Garra*, given to one species, and applied by me to the whole, seems to be derived from this circumstance. The fishes of this division have a considerable resemblance to the *Cyprinus aphya.*

80th Species.—CYPRINUS LAMTA.

A *Cyprinus* of the *Garra* kind, with four tendrils; with the pectoral fins sharp in the middle; with ten rays in the dorsal, and seven in the anal fin; and with the lateral line above the middle.

The *Lamta* I have found in rivulets, with rocky bottoms, in the province of Behar, and in the Rapti river of the Gorakhpur district. It grows to about three inches in length, and out of water retains life for a considerable time. The *colours* are green above, and silver below, with a faint spot on each side toward the end of the tail. The eyes are like silver.

The *head* is oval, blunt, and of moderate size. The *tendrils* are very minute. The nose projects a little beyond the mouth, which is small. The *jaws* protrude in opening. Below the under one is a circular space, callous in the middle. The *eyes* are small and high. The *membranes* of the gill-covers are exposed to view.

The ridge of the *back* is convex. The *lateral line* runs straight, near the middle of the side, but rises a little at the shoulder. The *scales* are large in proportion, and adhere firmly.

The *dorsal* fin is before the middle; the first of its rays is short, and, with the second, is undivided; the others are branched. The *pectoral* fins are nearly as long as the head; each contains thirteen rays. Each *ventral* fin has nine rays. The rays of the *anal* fin are like those on the back. The *tail* fin ends in a crescent, and contains nineteen distinct rays, besides short compacted ones.

81st Species.—Cyprinus Sada.

A *Cyprinus* of the *Garra* kind, with four tendrils; with the pectoral fins sharp above; with ten rays in the dorsal, and seven in the anal fin; and with the lateral line above the middle of the side.

This little fish I found in the Brahmaputra river, during the spawning season, on which account, perhaps, its belly appeared more prominent than usual. The *colours* are green above, and silver below, and this colour extends to the eyes.

The *head* is half oval, being flattened above, blunt, and of moderate size. The *nose* projects a little beyond the *mouth*, which is small. The *jaws* protrude in opening. The *tendrils* are shorter than the head. The *nostrils* are in the middle between the eyes and nose. The *membranes* of the gill-covers are exposed to view.

The *back* is nearly straight, with a convex edge. The *late-*

ral lines are straight. The *vent* is behind the middle. The *scales* are of moderate size.

The *back* fin is before the middle; the two first of its rays are undivided, and not indented; the others are branched. The *pectoral* fins are much smaller than the head, and low. Each *ventral* fin has nine rays. The three first rays of the *anal* fin are undivided and united together, the second being longer than the first, and shorter than the third; the others are branched. The *tail* fin is divided into two lobes.

82d Species.—Cyprinus latius.

A *Cyprinus* of the *Garra* kind, with two tendrils; and with eleven rays in the dorsal, and seven in the anal fin.

This small fish, which is tenacious of life out of water, I found in the Tista river. The *colour* above is greenish, with many dots, below silver. The lower fins are stained yellow. The eyes are like the body in colour, with a narrow golden circle round the pupil.

The *head* is oval, bluntish, and short, as are the *tendrils* proceeding from the sides of the *nose*, which projects beyond the *mouth*. This is small, and opens transversely. The *jaws* do not protrude in opening, and are covered with fleshy *lips*, of which the upper is bent inward, and indented on the edge. The *nostrils* are in the middle, between the eyes and end of the nose. The *eyes* are high and large, and have oval pupils, with the longer axis vertical. The *gill-covers* are dotted, and conceal half their *membranes*.

The *back* slopes a little before the fin, and has a convex ridge. The *lateral line* runs straight along the middle of the side. The *vent* is behind the middle. The *scales* are large. Above each ventral fin is a scale-like *appendant*.

The *dorsal* fin is near the middle; the three first of its rays are undivided, and smooth-edged, the second being longer than the first, and shorter than the third; the others are branched. The *pectoral* fins are shorter than the head, and low; each has about thirteen rays. Each *ventral* fin has nine rays. The *anal* fin, occupying the middle of the tail, slopes backward: the first of its rays is short, and closely united with the second, which, like it, is undivided; the others are branched. The *tail* fin is divided into two lobes, of which the upper is the longest; it contains twenty distinct rays, besides several compacted ones.

83d Species.—Cyprinus gohama.

A *Cyprinus* of the *Garra* kind? with two tendrils; and with eleven rays in the dorsal, and eight in the anal fin.

This little fish I have found both in the Kosi and Ganges. It differs very little from the *Latius*, but it dies very soon out of water, and its *form* is somewhat deeper and more protuberant on the back. The lower fins are pellucid like glass. The *pupils* are circular. The *lateral line* descends from the shoulder to the tail. Before the seven rays of the anal fin is an eighth minute one. The under jaw along its middle has a slight ridge, on which account it might, perhaps, be placed in the third division. In other respects it agrees entirely with the account given of the *Latius*.

84th Species.—Cyprinus mosario.

A *Cyprinus* of the *Garra* kind, without tendrils; with the pectoral fins sharp above; and with the sides free from cloud-like spots.

This is another small fish, differing very little from the *Latius*, and found in the Rapti river of the Gorakhpur district. It differs from the description of the *Latius* in wanting the tendrils, and the first ray of the dorsal fin.

85th Species.—Cyprinus sucatio.

A *Cyprinus* of the *Garra* kind, without tendrils; with the pectoral fins rounded; with the sides clouded with dots; and with nine rays in the dorsal, and seven in the anal fin.

This fish, both in form and colour, has a strong resemblance to many of the genus *Cobitis,* but wants tendrils. It is found in the rivers of Northern Bengal, and grows to about three inches long. The back is rather protuberant. Above the *colour* is greenish, with scattered dots; on the sides these are collected into clouds, and below the body is whitish and diaphanous. The fins of the back, breast, and tail, are dotted. The eyes are brown, with a narrow golden circle round the pupil.

The *head* is blunt, short, depressed, dotted, and as thick as the body. The *nose* extends beyond the *mouth,* which is small and transverse. The *jaws* protrude in opening, and have fleshy *lips.* The *eyes* are far back on the sides of the head, and are globular. The *apertures* of the gills are much smaller than usual in this genus. I cannot observe any rays in the *membranes* of the gill-covers.

The *back* is more prominent than the belly, and has a convex ridge. The *belly* is flat. The *tail* ends almost in a point. The *lateral line* runs straight below the middle of the side. The *vent* is nearly in the middle. The *scales* are large, adhere firmly, and are marked with lines diverging from the centre.

The *dorsal* fin is before the middle, slopes backward, and contains nine rays, of which the first is short, and, like the second, is undivided; the others are branched, and the last is divided to the root. The *pectoral* fins are near each other, and rounded; each contains thirteen rays. The *ventral* fins are as large as the pectorals, and are rounded; each contains nine rays, of which the first adheres to the second, and, like it, is undivided, as are also the third and ninth. The *anal* fin is nearer the end of the tail than the vent, and has seven rays like those on the back. The *tail* fin ends in a crescent, and has about sixteen distinct rays, besides small compacted ones.

86th Species.—CYPRINUS BALITORA.

A *Cyprinus* of the *Garra* kind, without tendrils; with the pectoral fins rounded; with a row of cloud-like spots on each side, and a third along the back; and with ten rays in the dorsal, and seven in the anal fin.

The *Balitora* I have found in the rivers towards the north-east of Bengal, where it grows to about two inches in length. The *form* tapers to both ends. The whole is diaphanous except the eyes, the gill-covers, the membrane lining the visceral cavity, and the spine, which are *coloured* like silver. The spots on the body consist of numerous black dots. The fins of the back and tail are spotted.

The *head* is blunt, short, depressed, dotted, and of moderate thickness. The *nose* projects beyond the *mouth*, which is small and transverse, and has fleshy *lips*. The *eyes* are far back on the sides of the head, and are small and globular. The *membrane* of each gill-cover, so far as I can perceive, has only two rays.

The ridge of the *back* is convex. The *belly* is flat. The *tail* is wedge-shaped. The *lateral lines* are straight. The *vent* is before the middle. The *scales* adhere firmly, and, for the size of the fish, are large; but are so diaphanous as to be scarcely distinguishable.

The *dorsal* fin is on the middle of the back, and slopes behind; the first of its rays is short, and, like the second, undivided; the others are branched, and the last divided to the root. Each *pectoral* fin has about twelve rays. The *ventral* fins, placed near the vent, have each nine rays. The *anal* fin is remote from the vent, and its rays are like those on the back. The *tail* fin is divided into two lobes, which contain about eighteen distinct rays, besides short compacted ones.

XXI. Genus.—Cobitis.

Fishes of the fifth order, with one dorsal fin, which is supported by rays, and which, with the pectoral fins, is unarmed; without teeth in the jaws; and with six or eight tendrils round the mouth.

All the fishes of this genus, that I know in India, are small; nor are any of them found in such a quantity as to render them of the least importance in an economical view. Those with cloud-like marks on their sides have the strongest resemblance to the *Loach* (*Cobitis barbatula*) both in appearance and habits, which need not, therefore, be described, as being perfectly known to naturalists. Those which are striped are more beautiful fishes, resembling the appearance of the others but slightly, and differing a good deal in habits, especially in swimming higher, and in not remaining so stationary on the bottom. In many species, beneath each eye, are one or two strong prickles, capable of being reflected into a slit behind. These, no doubt, serve as weapons of defence.

Besides the above mentioned circumstances, all the species of *Cobitis*, that I have described, have the following in common:

The *form* is long in proportion to the width. The *head* is small, and destitute of scales. The *tendrils* are shorter than the head, but thick, and sharp-pointed. The *nose* projects beyond the *mouth*, which is small, and has fleshy *lips*. The *eyes* are small, and high on the sides of the head. The *gill-covers* are flexible, and the *apertures* are small, or at least of moderate dimensions. There are not above three rays in each *membrane* of the gill-covers; and in several species there are either none, or, on account of their slenderness, they are indistinctly visible. The *vent* is behind the middle. There are seven distinct fins, those on the *back* and behind the *vent* being short.

1st Species.—Cobitis botia.

A *Cobitis* with a prickle under each eye; with six tendrils; with cloud-like marks on the sides; and with fourteen rays in the dorsal, and eight in each ventral fin.

The *Botia* is found in the rivers of the north-eastern parts of Bengal, and is nearly of the same size and qualities for eating as the *Loach*.

The *form* is compressed, and wedge-shaped. The *body* is slippery and covered with scales. On each side is a longitudinal silver *coloured* stripe, above which is olive green, clouded with spots of a deeper shade. Below the line is somewhat diaphanous. The head is reddish, with yellowish gill-covers. The fins are reddish and spotted.

The *head* is blunt. The prickle below each eye is strong and undivided. Four of the *tendrils* proceed from the upper lip, and two from the corners of the mouth. The *gill-covers* are smooth, and expose part of their *membranes*, each of which has about three rays; but they are not to be seen distinctly.

The *scales* are small. There is no *lateral line*.

The *dorsal* fin is near the middle: the three first of its rays are undivided, the second being longer than the first, and shorter than the third; the others are divided twice into two. The *pectoral* fins end in a point, and each has about thirteen rays. In the *anal* fin are eight rays, of which the first is short, and united to the second, both being undivided; the others are branched. The *tail* fin is large, fan-shaped, and rounded behind, and contains fifteen distinct rays, besides some shorter.

2d Species.—Cobitis gongota.

A *Cobitis* with two prickles beneath each eye; with six tendrils; with cloud-like marks on the sides; and with eleven rays in the dorsal, and six in each ventral fin.

This fish grows to six inches in length, and is found in the rivers of Northern Bengal towards the mountains. The *form* is slightly compressed. The *colour* is white on the sides, above clouded with dotted spots, and below silvery. The dorsal and tail fins are transversely striped. The pectoral fins are whitish.

The *head* is half oval, sloping below, and narrower than the body. The *prickles* beneath the eyes are short, strong, and united by a membrane. The *tendrils* are at equal distances from each other; four proceed from the upper lip, and two from the corners of the *mouth*, which is circular. The *jaws* protrude much in opening. The under *lip* is reflected at a sharp angle. The *nostrils* are small, and in the middle between the nose and eyes, which are far back and near each other, and have the longest axis of the pupil vertical. The membranes of the gill-covers are devoid of visible rays.

The *lateral line* runs straight above the middle. The *scales* are small, angular, and adhere firmly. Above each ventral fin is a scale-like *appendant.*

The *dorsal* fin is far back: the first of its rays is minute, and the second short, and both, together with the third, are undivided; the others are branched, and the last of them divided to the root. The *pectoral* fins are rounded, and each has ten rays. The *ventral* fins are smaller than the pectorals. The *anal* fin, in the middle between the vent and end of the tail, has eight rays like those of the back. The *tail* fin is rounded, and has seventeen rays.

3d Species.—Cobitis cucura.

A *Cobitis* with prickles below the eyes; with six tendrils; with cloud-like marks on the body; and with nine rays in the dorsal, and six in each ventral fin.

I found this fish in the river Kosi, where it seldom reaches three inches in length. The *form* is compressed. The sides are of a white *colour*, below is somewhat diaphanous, above is dotted. The back and sides are clouded with five irregular transverse dark bars, alternating with an equal number of large irregular dark spots.

The *head* is narrower than the body, and blunt. Beneath each eye are two short *prickles*. The under *jaw* is narrow. The *nostrils* are in the middle, between the nose and *eyes*, which are near each other, and have the longest axis of their pupils vertical. The *gill-covers* are dotted, and expose to view their *membranes*, in which no rays are discernible.

The *lateral line* runs straight above the middle of the side. The whole skin is devoid of distinguishable *scales*, nor above the ventral fin is there any *appendant*.

The *dorsal* fin is behind the middle: the two first of its rays are short and undivided, and the others are branched. The *pectoral* fins are much smaller than the head, and are low down: each contains about nine rays. The interior rays of each ventral fin are longer than the exterior, which I do not recollect having seen in any other fish. There are seven rays in the *anal* fin, and fifteen in that of the *tail*, which is rounded.

4th Species.—Cobitis guntea.

A *Cobitis* with prickles under the eyes; with six tendrils; with cloud-like spots on the sides; and with eight rays in the dorsal, and seven in each ventral fin.

The *Guntea* (*Gunté*) is found in the ponds and fresh rivers of Bengal, and grows to three or four inches in length. The *form* is compressed, and wedge-shaped. The skin is slimy. The body above is olive *coloured*, with a coppery stripe along each side, and below this is somewhat diaphanous. The gill-covers are greenish. The sides are clouded with dark coloured dots, collected into clusters, of which, on the middle of each side, there is a row more remarkable than the others. Above the end of each lateral line is a black spot. The rays of the back and tail fins are spotted with olive and white; those of the other fins with black and brazen.

The *head* is oblong, compressed, and blunt. Beneath each eye is a strong sharp *prickle*, divided at the end into two. Two *tendrils* proceed from the upper lip, and four from the angles of the mouth. The upper *jaw* is the longest. The *eyes* can be turned down so as to see the bottom. The *gill-covers* are rounded behind, and dotted; each of their *membranes* seems to contain three rays, but I cannot be certain of their number.

The *back* and *belly* are straight, and blunt-edged. There are no *lateral lines*. The body is covered with minute *scales*.

The first and second rays of the *dorsal* fin are undivided and closely united; the others are divided into two, the last of them to the root. Each of the *pectoral* fins contains eight rays. The *anal* fin contains seven rays like those of the back. The *caudal* fin ends in a straight line, and has seventeen rays.

5th Species.—COBITIS DARIO. Plate XXIX. Fig. 95.

A *Cobitis* with prickles under the eyes; with six tendrils; and with transverse bars on the sides.

The *Dario* (*Dari*) is found in the northern rivers of Bengal, and grows to two or three inches in length. The *form* is more prominent above than below, and compressed. The *colours* are disposed in transverse bands, slanting obliquely backward along the sides, and alternately yellow and covered with black dots. All the fins are free from marks, except that of the tail, which is yellow, with transverse black imperfect bands. The eyes are yellow, dotted with black.

The *head* is half oval, sloping above, and no wider than the body. Under each eye are two short strong *prickles*, joined by a membrane. Two of the *tendrils* hang from the corners of the mouth, and four from the end of the nose. The *mouth* is circular, with *jaws*, which protrude in opening. The *nostrils* are in the middle, between the eyes and nose, and each has two small apertures. The *eyes* are far back, and flat. The *gill-covers* are incapable of much motion; nor can I distinguish any rays in their *membranes*, which are thick.

The *back* is arched. The *lateral lines* run straight along the middle of the sides. The body is devoid of *scales*.

The *dorsal* fin is near the middle of the back, and has eleven rays, all nearly of the same length. The *pectoral* fins are low, much smaller than the head, and sharp above: each contains thirteen rays. The *ventral* fins are nearly as large as the pectorals, and near each other; and each has eight rays. The *anal* fin under the middle of the tail has seven rays, the first of which is short, and the others of nearly equal lengths. The *tail* fin is divided into two sharp lobes, and has eighteen distinct rays, besides some short compacted ones.

6th Species.—Cobitis pangia.

A *Cobitis* with prickles under the eyes; with eight tendrils; and without spots on the body.

The *Pangia (Pangya)* is found in the north-eastern parts of Bengal, grows to three or four inches in length, and is not eaten by the natives. The *form* is nearly of equal height throughout, and compressed. The fish is slippery, devoid of scales, and of a reddish *colour*, with dark dots, and its muscles are visible through the skin.

The *head* is oval, sloping at the nose, and narrow. The *prickle* below each eye is two-forked. Two *tendrils* proceed from the end of the nose, two from the corners of the mouth, and four from under the chin. The *nostrils* are in the middle between the nose and eyes, and are tubular. The *eyes* are minute. The *membranes* of the gill-covers contain no rays.

The *lateral lines* run straight above the middle of the sides. The *back* is straight.

The number of rays in the fins, owing to their slenderness, and to the thickness of the connecting membranes, cannot be exactly ascertained. The *dorsal* fin is behind the middle of the back, is rounded, and contains about seven rays. The *pectoral* fins are smaller than the head, and rounded; each contains about ten rays. The *ventral* fins are near each other, but do not extend to the vent; each has six rays. The *anal* fin is rounded, and has about seven rays. The *caudal* fin is small and rounded, and has about eighteen rays.

7th Species.—Cobitis geto. Plate XI. Fig. 96.

A *Cobitis* with prickles below the eyes; with eight tendrils; and with the body transversely barred.

The *Geto* is found in ponds in the north-eastern parts of Bengal. The *form* is compressed, and more prominent above than below. The whole is devoid of scales, and of a yellow *colour*, with transverse black bars running obliquely backward on the sides. One of the bars passes through the eye, which is black, with a gold-coloured ring round the pupil. The fins are yellow, with two black spots on that of the tail.

The *head* is half oval, sloping downward above, sharp, and not wider than the body. The *prickle* below each eye is divided to the root into two. Two *tendrils* proceed from the corners of the mouth, four from the end of the nose, and two from under the chin. The *mouth* is circular, with *jaws* that protrude in opening, the under one ending in a hard point. Each *nostril* has two apertures. Each *membrane* of the gill-covers has three rays.

The *back* is arched and round-edged. There is no *lateral line*.

The fin of the *back*, which is near its middle, and slopes behind, has twelve rays. The *pectoral* fins are much smaller than the head, and rounded: each contains thirteen rays. The *ventral* fins are rather larger, and each has nine rays. The *anal* fin has seven. The fin of the *tail* is divided into two sharp lobes.

8th Species.—COBITIS BALGARA.

A *Cobitis* with prickles below the eyes; with eight tendrils; and with the body spotted.

The *Balgara* is found in the Kosi river, and grows to about two inches in length. The *form* is a little compressed, and sharp before; and the tail is like a wedge. The whole is devoid of scales, and above is of a pale olive *colour*, with many darker dotted spots, and below is silvery. The spots on the back are oblong, and transverse; on the sides they are irregular. The fins of the back and tail are variegated.

The *head* is half oval, sloping above, and narrower than the body. The *prickle* below each eye is divided to the root. Two *tendrils* proceed from the corners of the mouth, four from the nose, and two from the chin. The *mouth* is small, and the *jaws* protrude much in opening. The *nostrils* are in the middle between the nose and eyes, and are either tubular, or have two apertures, separated by a projecting membrane; but are so small, that I am not certain which is their structure. The *eyes* are near each other, and have circular pupils. I cannot distinguish the number of rays in the *membranes* of the gill-covers.

The *dorsal* and *anal* fins, as well as both *pectoral* and *ventrals*, have each seven rays. The fin of the *tail* has seventeen, and ends in a crescent.

9th Species.—Cobitis savona.

A *Cobitis* with the head unarmed; with six tendrils; and with half bars on the sides.

The *Savona (Savon khorka)* is found in the Kosi river, where it grows to about two inches long.

The *form* before is sharp, behind only compressed. The whole is devoid of scales. Above the *colour* is dusky, with narrow yellowish bars; below it is white. The fin of the tail is dotted. The eyes are golden.

The *head* is half oval, sloping above. Four *tendrils* proceed from the upper jaw, and two from the corners of the mouth. The *nose* projects beyond the mouth. The *gill-covers* are shining.

The *back* rises to the fin with a sharp edge. The *lateral line* is straight and smooth.

The *back* fin has ten rays, of which the two first and the last are undivided. The *pectoral* fins are low, and smaller than the head: each contains about ten rays. Each *ventral* fin has about six rays, which is the number in the fin behind the *vent*, where the two first are undivided. The *tail* fin ends in a crescent.

10th Species.—Cobitis turio.

A *Cobitis* with the head unarmed; with six tendrils; with the sides clouded; and with eight rays in the dorsal fin, and also in each ventral.

The *Turio (Turi)* I found in the Brahmaputra river, where it grows two or three inches long.

The *form* is compressed, with a prominent back. The whole is devoid of scales, above silver *coloured*, clouded with dotted irregular spots, and below somewhat diaphanous. The eyes are blackish, with a golden ring round the pupil. The dorsal fin is yellow and spotted; the anal has no spots; the caudal is yellow, with transverse bars, and a black spot above near the root.

The *head* is half oval, sloping above, not wider than the body, and of a moderate size. Two *tendrils* proceed from the corners of the mouth, and four from the upper jaw. The *jaws* protrude a little in opening. The *eyes* are far back. The *gill-covers* have little motion. The rays of their *membranes* are to be seen very indistinctly.

The *back* is arched. The *lateral line* runs straight along the middle of the side.

The *dorsal* fin is before the middle of the back: the three first of its rays are closely united, the first being very short. The *pectoral* fins are sharpish in the middle; each has about twelve rays. The *ventral* fins are smaller, and the *anal* has seven rays. The *tail* fin ends in a notch, and has nineteen rays.

11th Species.—Cobitis bilturio.

A *Cobitis* with the head unarmed; with six tendrils; with clouded sides; and with fourteen rays in the dorsal, and eight in each of the ventral fins.

The *Bilturio (Bil turi)* I found along with the tenth species, from which it differs scarcely in any thing but in the number of rays which support the fins. Each *pectoral* fin has fourteen rays; that behind the *vent* has seven.

12th Species.—Cobitis corica.

A *Cobitis* with the head unarmed; with six tendrils; with the body spotted; and with nine rays in the dorsal fin.

The *Corica (Khorika)* is found in the river Kosi, where it grows to about two inches in length.

The *body* is not compressed, but is sharp before, and has a wedge-like tail. The whole is devoid of scales. The upper parts are of a bluish *colour*, and somewhat diaphanous, with many large spots, composed of black dots. The lower parts are silvery, as are the eyes.

The *head* slopes above, and is half oval. Four *tendrils* proceed from the upper jaw, of which the two in the centre are very minute: the other two proceed from the corners of the mouth. The *nostrils* are near the *eyes*, which are far back. The *gill-covers* are shining and dotted. Their *membranes* are half exposed to view; but their rays, on account of their slenderness, cannot be exactly numbered.

On the *back* is a longitudinal furrow. The *lateral line* forms a ridge on the middle of the side.

The *dorsal* fin is on the middle of the back: the first of its rays is undivided, and the last divided to the root. The *pectoral* fins are longer than the head, and sharp above: each has eleven rays, of which the two first are short, and the third very long and undivided. The *ventral* fins are smaller, and each has seven rays; in the *anal* there are six. The *caudal* fin is divided into two blunt lobes, and has nineteen distinct rays, besides short compacted ones.

INDEX METHODICUS.

In hoc indice L. B. notandum est, quod in radios pinnarum enumerando notam post aliquem numerum interrogationis (?) significare numerum vel ob cutis crassitudinem, vel ob radiorum tenuitatem esse incertum ; notam autem (—) significare me hujus pinnæ radios non percensuisse ; notam denique (+) significare radios præter annotatos esse nonnullos ita arcte compactos ut eorum numerus incertus restat.

ORDO I.—PISCES CARTILAGINEI.

Pisces spina dorsali cartilaginea.

Genus I. Raia. Spiracula subtus corpus utrinque 5 s. 7, operculis et membranis branchiostegis nudata ; os infra rostrum ; corpus depressissimum, pinnis pectoralibus abdominalibusque marginatum.

Species 1. *Raia Lymma ?* cauda elongata, spina caudali serrata.

2. *Raia Sancur* dentibus obtusis, corpore caudaque pinnata inermibus, dorso apterygio, capite circinnato.

Genus II. Squalus. Spiracula in colli latera utrinque 5 seu 7, operculis et membranis branchiostegis nudata.

Species 1. *Squalus Carcharias ?* pinna anali solitaria ; apertura auriculari nulla ; dentibus triangularibus, utrinque serratis ; pinna caudali caudæ versus finem subtus deficiente.

2. *Squalus pectinatus* rostro elongato, ancipite, spinis angustioribus utrinque circiter 34 armato.

Genus III. Tetrodon maxillis porrectis, osseis, bifidis ; corpore ad pisces arbitrium inflato, alepidoto.

Species 1. *Tetrodon fluviatilis* corpore vix compressiusculo, supra fusco-virdi, lineis flavescentibus decussantibus in pulvinullos quasi diviso; maxilla superiore sublongiore.

B 1? D 16, P 17, A 14, C 8.

2. *Tetrodon Patoca* corpore vix compressiusculo, spinulis simplicibus hispido, supra nigricante maculis flavis vario; maxilla superiore longiore.

B—? D 11, P 16? A 10, C 10.

3. *Tetrodon Cutcutia* corpore vix compressiusculo, inermi, supra viridi plagis irregularibus fuscis, infra albido; macula magna nigra utrinque ad caudæ partem anteriorem; maxilla inferiore longiore.

B 0, D 12, P 21? A 11, C 8.

4. *Tetrodon Caria* corpore vix compressiusculo, inermi, supra plagis irregularibus fuscis viridi, infra nigro; maxillis subæqualibus.

B—, D 11, P 18? A 10? C 8.

5. *Tetrodon Gularis* corpore vix compressiusculo, inermi, supra plagis irregularibus fuscis viridi, infra nigro; macula utrinque nigra ad caudæ partem anteriorem; plaga gulari rubra; maxillis subæqualibus.

6. *Tetrodon Tepa* corpore compresso, aculeato, inflationis capace; pinna caudæ bilobo-lunata.

B—, D 14, P 14, A 11, C 13.

Genus IV. SYNGNATHUS oris rictu ad apicem rostri longissimi subcilindrici minuto, edentulo; branchiarum apertura in nucha, per membranam adnatam sæpius obturata.

Species 1. *Syngnathus Cuncalus* pinnis dorsi pectoralibus et caudæ radiatis; corpore septemangulari; cingulis abdominalibus 16, caudalibus 25.

B 2, D 50, P 16, C 9.

2. *Syngnathus Carce* pinnis dorsi pectoralibus et caudæ radiatis; corpore septemangulari; cingulis abdominalibus 32, analibus 18.

B—, D 25, P 10, C 10.

Species 3. *Syngnathus Deocata* pinnis dorsi pectoralibus et caudæ radiatis; corpore sexangulare.

B—, D 30? P 15? C 7?

ORDO II.—PISCES APODES.

Pisces spina dorsali ossea absque pinnis ventralibus.

Genus I. UNIBRANCHAPERTURA pinnis pectoralibus nullis; corpore tereti serpentiformi; apertura branchiarum unica, gulari.

Species 1. *Unibranchapertura Cuchia* pinnis nullis; corporis maculis atris.

Genus II. MURÆNOPHIS corpore serpentiformi, alepidoto; pinnis pectoralibus nullis, dorsi caudæ anique unitis.

Species 1. *Murænophis Sathete* pinna dorsi e nucha incipiente; dentibus in utraque maxilla acutis, remotis; maxillis subæqualibus; corpore emaculato.

2. *Murænophis Tile* pinna dorsi e nucha incipiente; dentibus in utraque maxilla acutis, remotis; dente in palato unico, magno; corpore lineolis depressis tessulato, et lituris flavis et punctis nigris variegato.

Genus III. OPHISURUS pinnis pectoralibus dorsi et ani radiatis, duabus posterioribus longitudinalibus, linearibus; cauda nuda; corpore serpentiformi; apertura branchiostega in utroque latere.

Species 1. *Ophisurus Hijala* maxilla superiori longiore; ordine macularum pallidarum prope lineæ lateralis partem anteriorem.

B 0, D 230-240, P 8, A 170?

2. *Ophisurus Boro* maxilla superiore longiore; corpore absque linea laterali, immaculato; radiis dorsalibus 330 circiter.

B 0, D 330? P 11? A 270?

3. *Ophisurus Harancha* maxilla superiore longiore; corpore immaculato; linea laterali recta; radiis dorsalibus 345 circiter.

B 0, D 345? P 12? A 240?

Genus IV. MURÆNA corpore serpentiformi, alepidoto, lubrico, pinnis pectoralibus ani caudæ dorsique instructo; dorso caudaque inermibus.

Species 1. *Muræna Anguilla* maxilla inferiore longiore; radiis analibus 100 circiter; pinnis ani caudæ dorsique unitis; corpore immaculato.

2. *Muræna maculata?* maxilla inferiore longiore; corpore viridi, nebulis nigris variegato; pinnis ani caudæ dorsique unitis.

P 24.

3. *Muræna Vamos* maxilla superiore multo longiore; corpore immaculato; linea laterali nulla; pinnis ani caudæ dorsique unitis.

4. *Muræna Bagio* maxilla superiore multo longiore; corpore supra bruneo, immaculato; linea laterali evidente; pinnis ani caudæ dorsique unitis.

D 260? A 220? P 12.

5. *Muræna Raitaborua* maxilla superiore longiore; corpore supra purpureo, punctis nigris asperso; pinnis ani caudæ dorsique distinctis.

B–, D 32, P 8, A 50, C 60.

Genus V. MACROGNATHUS corpore squamoso, subserpentiformi, ante pinnas dorsi et ani aculeato; cauda pinnata.

Species 1. *Macrognathus armatus* pinnis dorsi caudæ anique unitis.

B—, D 83—85, P 21, A 88—90, C 17.

2. *Macrognathus aculeatus* pinna caudæ distincta; ocellis in pinna dorsi 3—7.

B 5? D 51? P 20, A 48? C 14.

3. *Macrognathus Pancalus* pinna caudæ distincta; corpore punctis nigris et maculis parvis pallidis variegato.

B 4, D 40? P 12, A 25? C—.

Genus VI. TRICHIURUS corpore compressissimo; cauda subulata, pinnis nudata.

Species 1. *Trichiurus Lepturus* maxilla inferiore longiore.

B—, D 114, P 11.

Genus VII. CYNOGLOSSUS oculis ad unum latus positis, corpore compressimo in latus oculis oppositum incumbente.

Species 1. *Cynoglossus Lingua* oculis sinistris; pinnis dorsi caudæ anique unitis, pectoralibus nullis.

B—, D 140? A 111, C 8.

ORDO III.—PISCES JUGULARES.

Pisces spina dorsali ossea, pinnis ventralibus ante pectorales positis.

Genus I. BATRACHOIDES capite magno depresso; oris rictu maximo; cirrhis circa maxillis nonnullis.

Species 1. *Batrachoides Gangene* corpore alepidoto, filamentis asperso; aculeis dorsalibus 3 recumbentibus, in operculo branchiarum 4.

B 4, D $\frac{3}{3}$ et 19, P 22, V $\frac{1}{6}$, A 16, C 14?

ORDO IV.—PISCES THORACICI.

Pisces spina dorsali ossea, pinnis ventralibus sub pectorales positis.

Genus I. GOBIOIDES pinnis ventralibus unitis; cauda pinnata; dorso monopterigio.

Species 1. *Gobioides rubicundus* pinna caudæ cum dorsali analique unita; radiis dorsalibus 46, analibus 40.

B 5, D 46, P—, V 12, A 40, C—.

2. *Gobioides ruber* pinna caudæ discreta; radiis dorsalibus 55, analibus 52.

B—, D 55, P 20, V 10, A 52, C 16.

Genus II. GOBIUS pinnis ventralibus unitis, dorsalibus duabus.

Species 1. *Gobius Bato* corpore anguilliformi; radiis dorsalibus 5 et 21, analibus 22.

B 3, D 5 et 21, P 21, V 12, A 22, C 22?

2. *Gobius Changua* corpore anguilliformi; radiis dorsalibus 5 et 31, analibus 29.

B 4, D 5 et 31, P 18, V 12, A 29, C 18.

Species 3. *Gobius viridis* corpore anguilliformi; radiis dorsalibus 5 et 26, analibus 26.

B 3, D 5 et 26, P 19, V 12, A 26, C 16.

4. *Gobius Plinianus* pinnis pectoralibus flabello insidentibus; radiis pinnæ dorsalis anterioris 5 elongatis, posterioris et ani utriusque 26.

B 3, D 5 et 26, P 17, V 12, A 26, C 17.

5. *Gobius septemradiatus* pinnis pectoralibus flabello insidentibus; radiis pinnæ dorsalis anterioris 7 abreviatis, posterioris 13, analis 14.

B 2, D 7 et 13, P 13, V 12, A 14, C 17.

6. *Gobius novemradiatus* pinnis pectoralibus flabello insidentibus; radiis pinnæ dorsalis anterioris 9, posterioris 14, analis 15.

B 0, D 9 et 14, P 13, V 12, A 15, C 13.

7. *Gobius tredecemradiatus* pinnis pectoralibus flabello insidentibus; radiis pinnæ dorsalis anterioris 13, posterioris 14, analis 12.

B 3, D 13 et 14, P 12, V 12, A 12, C 14+.

8. *Gobius Gutum* corpore cuneato; radiis pinnæ dorsalis posterioris 11, prioris radios longitudine equantibus; radiis pectoralibus 13; corpore diaphano, punctis nigris, nebuloso.

B 3, D 6 et 11, P 13, V 10, A 11, C 17.

9. *Gobius Giuris* corpore cuneato; radiis pinnæ dorsalis posterioris 10, pectoralium 22, analis 9.

B 5, D 6 et 10, P 22, V 12, A 9, C 14+.

10. *Gobius Sadanundio* corpore cuneato; radiis pinnæ dorsalis posterioris 8.

B—, D 6 et 8, P 18, V 9, A 9, C 23.

11. *Gobius Chuno* corpore cuneato; radiis pinnæ dorsalis posterioris 7.

B—, D 5 et 7, P 10, V 5, A 10, C 18?

12. *Gobius Nunus* corpore cuneato; radiis pinnæ dorsalis posterioris 5.

B—, D 6 et 5, P 10, V 10, A 7, C 13?

Genus III. CHEILODIPTERUS pinnis ventralibus distinctis, dorsalibus duabus; operculis branchiostegis integerrimis, inermibus; apertura branchiarum utrinque ampla.

Species 1. *Cheilodipterus Culius* pinna caudali rotundata; radiis pinnæ dorsalis anterioris 6 inermibus.

B 4, D 6 et 10, P 17, V 6, A 11, C 27.

2. *Cheilodipterus Butis* pinna caudali rotundata; aculeis pinnæ dorsalis anterioris 6.

B 7, D $\frac{6}{6}$ et 9, P 18, V 5, A 9, C 15?

3. *Cheilodipterus Panijus* pinna caudali subtruncata; aculeis pinnæ dorsalis anterioris 9, quorum secundus longissimus.

B 5, D $\frac{9}{9}$ et 28, P 20, V $\frac{1}{6}$, A 28, C 19+.

Genus IV. OPHIOCEPHALUS pinnis inermibus, dorsali unico, ventralibus geminis; capite operculisque branchiostegis inermibus, integerrimis, et squamosis; corpore elongato, parum compresso.

Species 1. *Ophiocephalus Wrahl?* radiis pinnæ dorsalis 43-46; fasciis pluribus nigris latera variegantibus; pinnis dorsi et ani pone nigro maculatis.

B 6, D 43—46, P 17, V 6, A 26—28, C 14+.

2. *Ophiocephalus Chena* radiis pinnæ dorsalis 43—46; dorso pinnis que dorsi et ani immaculatis.

B 6, D 43—46, P 17, V 6, A 26—28, C 14+.

3. *Ophiocephalus Lata* radiis pinnæ dorsalis 30 circiter; dorso fasciis atris pluribus ad vittam longitudinalem atram descendentibus variegato.

B 5, D 30? P 16, V 6, A 20? C 12+.

4. *Ophiocephalus Marulius* radiis pinnæ dorsalis 52—54, ventralium senis; macula pinnæ caudalis ocellata.

B 4, D 52—54, P 19, V 6, A 31—35, C 15.

5. *Ophiocephalus Barca* radiis pinnæ dorsalis 52 circiter, ventralium quinis.

B 4, D 52? P 16, V 5, A 35, C 19.

6. *Ophiocephalus Gachua* radiis pinnæ dorsalis 36 circiter, ventralium quinis; fasciis dorsi nigris plurimis.

B 5, D 36? P 15, V 5, A 22? C 12?

Species 7. *Ophiocephalus aurantiacus* radiis pinnæ dorsalis 34 circiter; colore ubique aurantiaco.

Genus V. LABRUS maxillis mobilibus; operculis branchiostegis inermibus, integerrimis; pinnarum nonnullis aculeatis, dorsali unico; corpore compresso.

Species 1. *Labrus Badis* radiis simplicibus recurvis 17, et compositis 8 in pinna dorsali; aculeis 3, et radiis compositis 7 in pinna anali.

B 5, D 25, P 12, V 6, A $\frac{3}{10}$, C 16.

2. *Labrus Dario* aculeis pinnæ dorsalis 14, radiis compositis 8; aculeis analibus 3, radiis compositis 7.

B—, D $\frac{14}{22}$, P 9, V 6, A $\frac{3}{10}$, C—.

Genus VI. BOLA dentibus vel conicis acutis, vel in ordines plures confertis; ossium capitis nonnullis serratis; pinnarum nonnullis aculeatis; corpore opaco; lateribus declivibus.

Species 1. *Bola Coitor* aculeis pinnæ dorsalis anterioris 10; aculeo 1 radiisque mollibus 26 vel 27 in pinnarum dorsalium posteriore; aculeis 2 et radiis mollibus 6 seu 7 in pinna anali; ramo primario radii pinnarum ventralium secundarii setaceo-producto; rostro ultra maxillas elongato.

B 7, D $\frac{10}{10}$ et $\frac{1}{27-28}$, P 17, V $\frac{1}{6}$, A $\frac{2}{8-9}$, C 17+.

2. *Bola Chaptis* aculeis pinnæ dorsalis anterioris 10; aculeo 1 radiisque mollibus 25 in pinnarum dorsalium posteriore; aculeis 2 et radiis mollibus 7 in pinna anali; ramo primario radii ventralium secundarii setaceo-producto; maxillis terminalibus.

B 7, D $\frac{10}{10}$ et $\frac{1}{26}$, P 19, V $\frac{1}{6}$, A $\frac{2}{9}$, C 17+.

Varietas Nuria dicta in pinna dorsali posteriore radios molles 24 habet, cæteroquin non diversa.

3. *Bola Coïbor* aculeis pinnæ dorsalis anterioris 9; aculeo 1 radiisque mollibus 24 in pinnarum dorsalium posteriore; aculeis 3 radiisque mollibus 7 in pinna anali; pinnis ventralibus imberbibus; ventre lateribusque flavis.

4. *Bola Pama* aculeis pinnæ dorsalis anterioris 9; aculeo 1

radiisque mollibus 41—45 in pinnarum dorsalium posteriore; aculeis 2 radiisque mollibus 7 in pinna anali; pinnis ventralibus imberbibus; maxillis subterminalibus; lateribus argenteis.

B 7, D $\frac{9}{9}$ et $\frac{1}{41-45}$, P 17, V $\frac{1}{6}$, A $\frac{2}{9}$, C 17 +.

Species 5. *Bola Cuja* aculeis pinnæ dorsalis anterioris 10; aculeo 1 radiisque mollibus 28 in pinnarum dorsalium posteriore; aculeis 2 radiisque mollibus 7 in pinna anali; ramo primario radii pinnarum ventralium secundarii setaceo producto; maxillis terminalibus, inferiore longiore; vittis laterum nigris, interruptis.

B 7, D $\frac{10}{10}$ et $\frac{1}{29}$, P 17, V $\frac{1}{6}$, A $\frac{2}{6}$, C 17 +.

6. *Bola? Coïoides* aculeis 11 radiisque mollibus 15 in pinnis dorsi unitis; aculeis 3 radiisque mollibus 8 in pinna anali; maculis laterum circinnatis nigris ferrugineisque.

B 7, D $\frac{11}{26}$, P 18, V $\frac{1}{6}$, A $\frac{3}{11}$, C 16+.

Genus VII. Coius dentibus in utraque maxilla vallidis acutis, vel in ordinibus pluribus congestis; ossium capitis nonnullis margine denticulatis; pinnarum nonnullis aculeatis; corpore opaco; lateribus subverticalibus.

Species 1. *Coïus Vacti* pinna caudali flabelliformi; pinnarum dorsalium subdistinctarum aculeis in priore 7, in posteriore aculeo 1 radiisque mollibus 11; aculeis in pinna anali 3, radiis mollibus 8.

B 7, D $\frac{7}{7}$ et $\frac{1}{12}$, P 17, V $\frac{1}{6}$, A $\frac{3}{11}$, C 18?

2. *Coïus Datnia* pinna caudali bifida; pinnarum dorsalium unitarum aculeis 12, et radiis mollibus 10; pinnæ analis aculeis 3, et radiis mollibus 8; ramo radii in pinnis ventralibus secundarii primario in setam producto.

B 6, D $\frac{12}{12}$ et 10, P 15, V $\frac{1}{6}$, A $\frac{3}{11}$, C 17+.

3. *Coïus Catus* pinna caudali sublunata; pinnarum dorsalium unitarum aculeis 10, radiis mollibus 14; pinnæ analis aculeis 3 radiis mollibus 8; pinnarum ventralium radii secundarii ramo primario in setam producto; macula diffusa nigra in latere utroque.

B 7, D $\frac{10}{10}$ et 14, P 17, V $\frac{1}{6}$, A $\frac{3}{11}$, C 17+.

Species 4. *Coïus trivittatus* pinna caudali bifurca; pinnarum dorsalium unitarum aculeis 11, radiis mollibus 10; pinnæ analis aculeis 3, radiis mollibus 8; vittis in utroque latere 3 nigris.

B 6, D $\frac{10}{10}$ et $\frac{1}{11}$, P 13, V $\frac{1}{6}$, A $\frac{3}{11}$, C 18 +.

5. *Coïus Gudgutia* pinna caudali truncata; pinnarum dorsalium unitarum aculeis 12, radiis mollibus 14; pinnæ analis aculeis 3, radiis mollibus 7; pinnarum ventralium radii secundarii ramo primo in setam producto.

B 7, D $\frac{11}{11}$ et $\frac{1}{15}$, P 17, V $\frac{1}{6}$, A $\frac{3}{10}$, C 17+.

6. *Coïus Polota* pinna caudali flabelliformi; pinnarum dorsalium unitarum aculeis 12, et radiis mollibus totidem; pinnæ analis aculeis 3, radiis mollibus 9; pinnarum ventralium radio secundario in setas duas producto.

B 6, D $\frac{12}{24}$, P 19, V $\frac{1}{6}$, A $\frac{3}{12}$, C 17+.

7. *Coïus Nandus* pinna caudæ flabelliformi; pinnarum dorsi unitarum aculeis 13, radiis mollibus 12; pinnæ analis aculeis 3, radiisque mollibus 8; maculis corporis nebulosis.

B 7, D $\frac{13}{25}$, P 15, V $\frac{1}{6}$, A $\frac{3}{11}$, C 16.

8. *Coïus Cobojius* pinna caudali subrotundata; pinnarum dorsi unitarum aculeis 17, etradiis mollibus 8; pinnæ analis aculeis 10, radiisque mollibus totidem; fasciis laterum nigricantibus obsoletis.

B 0? D $\frac{17}{25}$, P 16, V $\frac{1}{6}$, A $\frac{10}{20}$, C 14?

9. *Coïus Chatareus* pinna caudali retusa; pinnæ dorsalis solitariæ aculeis 5, radiis mollibus 12; pinnæ analis aculeis 3, radiis mollibus 16; corpore argenteo; maculis dorsi nigris.

B 7, D $\frac{5}{17}$, P 12, V $\frac{1}{6}$, A $\frac{3}{19}$, C 17+.

Genus VIII. Chanda dentibus in utraque maxilla acutis, sparsis, vel obsoletis; margine ossium in capite nonnullorum inciso; pinnarum nonnullis aculeatis; corpore plus minus diaphano, compresso; lateribus subverticalibus.

Species 1. *Chanda? setifer* pinnæ dorsalis solitariæ aculeis 10, radiisque mollibus 9; analis aculeis 3, radiisque mollibus 7; ramo primario radii in pinnis ventralibus secundarii in setam producto; corpore squamoso, subopaco.

B 6, D $\frac{10}{19}$, P 16, V $\frac{1}{6}$, A $\frac{3}{10}$, C 17+.

Species 2. *Chanda Ruconius* pinnæ dorsalis solitariæ aculeis 8, radiisque mollibus 16; pinnæ analis aculeis 3, radiisque mollibus 14; corpore squamoso, subdiaphano, argenteo; fasciis dorsi nigris punctatis.

B 6, D $\frac{8}{24}$, P 16, V $\frac{1}{6}$, A $\frac{3}{17}$, C 17+.

3. *Chanda Nalua* pinnis dorsalibus 2 unitis, posterioris radiis mollibus 11; pinnæ analis radiis mollibus 10; capite alepidoto, corpore squamoso.

B 6, D $\frac{7}{7}$ et $\frac{1}{12}$, P 13, V $\frac{1}{6}$, A $\frac{3}{13}$, C 15+.

4. *Chanda Nama* pinnis dorsalibus unitis, posterioris radiis mollibus 16; pinnæ analis radiis mollibus 15; corpore elongato, alepidoto.

B 6, D $\frac{7}{7}$ et $\frac{1}{17}$, P 11, V $\frac{1}{6}$, A $\frac{3}{18}$, C 17+.

5. *Chanda Phula* pinnis dorsalibus unitis, posterioris radiis mollibus 14; pinnæ analis radiis mollibus 13; corpore elongato, alepidoto.

B 6? D $\frac{7}{7}$ et $\frac{1}{15}$, P 10 vel 11, V $\frac{1}{6}$, A $\frac{3}{16}$, C 17+.

6. *Chanda Bogoda* pinnis dorsalibus unitis, posterioris radiis mollibus 16; pinnæ analis radiis mollibus 17; corpore elongato, alepidoto.

B 6, D $\frac{7}{7}$ et $\frac{1}{17}$, P 12, V $\frac{1}{6}$, A $\frac{3}{20}$, C 16+.

7. *Chanda Baculis* pinnis dorsalibus unitis, posterioris radiis mollibus 13; pinnæ analis radiis mollibus 13; corpore brevi, alepidoto; macula nuchæ flava.

B 6, D $\frac{7}{7}$ et $\frac{1}{14}$, P 10? V $\frac{1}{6}$, A $\frac{3}{16}$, C 17+.

8. *Chanda Ranga* pinnis dorsalibus unitis, priore incisa, posterioris radiis mollibus 12; pinnæ analis radiis mollibus 15; corpore brevi, squamoso.

B 6, D $\frac{7}{7}$ et $\frac{1}{13}$, P 11, V $\frac{1}{6}$, A $\frac{3}{18}$, C 18+.

9. *Chanda Lala* pinnis dorsalibus unitis, priore integerrima,

posterioris radiis mollibus 12-14; pinnæ analis radiis mollibus 15; corpore brevi, squamoso.

B 6, D $\frac{7}{7}$ et $\frac{1}{15-15}$, P 11, V $\frac{1}{6}$, A $\frac{5}{18}$, C 18+.

Genus IX. Trichopodus pinnis ventralibus setaceis, uniradiatis.

Species 1. *Trichopodus Colisa* pinna caudali flabelliformi, integra; lateribus viride æreque fasciatis; pinna dorsi pone rotundata.

B—, D $\frac{16}{27}$, P 10, V 1, A $\frac{17}{34}$, C 18.

2. *Trichopodus Bejeus* pinna caudali emarginata; lateribus viridi cupreoque fasciatis; pinna dorsi pone acuta.

B 3? D $\frac{17}{26}$, P 9, V 1, A $\frac{16}{33}$, C 18.

3. *Trichopodus Cotra* pinna caudali flabelliformi, integra; lateribus viridi cæruleoque fasciatis.

B 4? D $\frac{17}{25}$, P 9, V 1, A $\frac{19}{35}$, C 17.

4. *Trichopodus Lalius* pinna caudali flabelliformi, integra; fasciis laterum rubris.

B 4? D $\frac{16}{24}$, P 9, V 1, A $\frac{18}{33}$, C 16+.

5. *Trichopodus Sota* pinna caudali retusa; coloribus corporis oblique demidiatis, supra et pore rufescente, infra et ante subargenteo.

B 4? D $\frac{18}{26}$, P 9, V 1, A $\frac{19}{30}$, C 16?

6. *Trichopodus Chuna* pinna caudali retusa; vitta aurea nigro punctata in latere utroque.

B—, D $\frac{17}{24}$, P 9, V 1, A $\frac{19}{30}$, C 15+.

Genus X. Chætodon ossibus capitis integerrimis; corpore compressissimo, brevi, squamoso; dentibus in utraque maxilla simplice serie positis, acerosis, approximatis, parallelis.

Species 1. *Chætodon Pairatalis* pinna caudali flabelliformi, integra; aculeis in pinnis dorsalibus subunitis 11, radiis mollibus 16; corpore maculato; ramo primario radii in pinnis ventralibus secundarii producto, subulato.

B 4, D $\frac{10}{10}$ et $\frac{1}{17}$, P 18, V $\frac{1}{6}$, A $\frac{3}{3}$ et $\frac{1}{15}$, C 14+.

2. *Chætodon Caris* pinna caudali retusa; aculeis dorsi 18, radiis mollibus 14; lateribus supra viridibus, subtus albidis; fasciis 5 latis, flavis.

B 5, D $\frac{18}{32}$, P 17, V $\frac{1}{6}$, A $\frac{13}{21}$, C 16+.

Genus XI. PLEURONECTES oculis utrisque in laterum altero dispositis; pectore utrinque pinnato.

Species 1. *Pleuronectes Nauphala* oculis sinistris; pinna caudali acuta; radiis dorsalibus 73, analibus 55; corpore elongato, tuberculis destituto.

B 7, D 73, P dext. 11, sinist. 12, V 6, A 55, C 17.

2. *Pleuronectes Arsius* oculis sinistris; pinna caudali integra; radiis dorsalibus 81, analibus 55; corpore elongato; linea laterali supra ocellis duobus signata.

B 3 s. 4, D 81, P 10, V 6, A 55, C 16.

3. *Pleuronectes Pan* oculis dextris; pinnis dorsi caudæ anique unitis; corpore brevi; radiis in pinna pectorali utraque 5 setaceis.

B 4, D C et A 115, supra 64 infra 51, P 5, V 4.

Genus XII. ACHIRUS oculis in altero laterum positis, pinnis pectoralibus nullis.

Species 1. *Achirus Cynoglossus* oculis sinistris; pinnarum dorsi caudæ anique unitarum radiis 102, 12, et 78; latere sinistro fusco punctato.

B 0, V 4, D C et A 102 et 12 et 78.

Genus XIII. CALLIOMORUS capite armato, ossibus integerrimis; corpore depresso, squamoso; pinnis dorsalibus duabus.

Species 1. *Calliomorus Chaka* spina ante oculum utrumque; lamina operculorum branchiostegorum posteriore inermi.

B 7, D $\frac{8}{8}$ et 13, P 18, V 6, A 13, C 18?

ORDO V.—PISCES ABDOMINALES.

Pisces spina dorsi ossea; pinnis ventralibus pone pectorales insidentibus.

Genus I. PLATYSTACUS cuti glutinosa, verrucosa, alepidota; ore cirrhis cincto; pinna dorsali unica, antica, brevi; cauda quam trunco et capite unitis breviore, subulata; pinnis dorsi pectorisque armatis.

Species 1. *Platystacus Chaca* pinna caudali supra ante analem, infra prope eandem extensa.

B 0, D $\frac{1}{4}$, P $\frac{1}{9}$? V 8, A 8, C 38?

Genus II. PLOTOSUS cute glutinosa, lævi, alepidota; ore cirrhis cincto; pinna dorsali postica cum caudali et anali in unem coalita; cauda quam capite truncoque unitis longiore.

Species 1. *Plotosus Canius* dorso unicolore; aculeo pinnæ dorsalis integerrimo.

Genus III. MACROPTERONOTUS cute lævi, glutinosa, alepidota; ore cirrhoso; pinna dorsali unica, elongata; cauda cuneata; pinnis pectoralibus armatis.

Species 1. *Macropteronotus Jagur* pinnis dorsi caudæque unitis.

B 2? D 53, P $\frac{1}{8}$, V 6, A 50, C 14?

2. *Macropteronotus Magur* pinnis dorsi caudæque distinctis, prioris radiis 60; radiis analibus 44; aculeis pectoralibus integerrimis.

B 9? D 60, P $\frac{1}{10}$, V 6, A 44, C 15.

Genus IV. SILURUS cute lævi, glutinosa, alepidota; ore cirrhoso; pinna dorsali unica, brevissima, scapulari, radiata; cauda longa, compressa; pinnis dorsalibus vel pectoralibus armatis, vel radiis branchiostegis plusquam tribus.

Species 1. *Silurus Singio* pinna caudali rotundata; cirrhis 8; aculeis pectoralibus utrinque retrorsum serratis.

B 0? D 7, P $\frac{1}{8}$, V 6, A 61? C 12+.

2. *Silurus Pabda* pinna caudali bifida, obtusa; cirrhis 4, quorum duo superiores pinnis pectoralibus longiores; radiis pinnæ analis 54; aculeis pinnarum pectoralium integerrimis; lateribus nigro nebulosis.

B 9, D 4, P $\frac{1}{12}$, V 8, A 54, C 17.

3. *Silurus Canio* pinna caudali bifida, acuta; cirrhis 4, quorum duo superiores ad piscis medium pertingunt; radiis pinnæ analis 69; aculeis pinnarum pectoralium integerrimis; lateribus immaculatis.

B 12? D 4, P $\frac{1}{12}$, V 8, A 69, C 18.

Species 4. *Silurus Duda* pinna caudali bifida, acuta; cirrhis 4, quorum 2 superiores ad piscis medium pertingunt; radiis pinnæ analis 73; aculeis pinnarum pectoralium integerrimis; lateribus immaculatis.

B 12? D 4, P $\frac{1}{12}$, V 8, A 73, C 18.

5. *Silurus Chechra* pinna caudali bifida, obtusa; cirrhis 4, quorum duo superiores ad pinnæ dorsalis apicem pertingunt; radiis pinnæ analis 67; aculeis pinnarum pectoralium pone denticulatis; lateribus nigro nebulosis.

B 13? D 4, P $\frac{1}{12}$, V 8, A 67, C 17.

6. *Silurus Pabo* pinna caudali biloba; cirrhis 4 capite brevioribus; radiis pinnæ analis 73.

B 13? D 5, P $\frac{1}{15}$, V 10, A 73? C 19.

7. *Silurus Boalis* pinna caudali biloba; cirrhis 4, quorum 2 superiores ad piscis medium pertingunt; pinnis pectoralibus inermibus; radio pinnæ dorsalis primario integerrimo.

B 19, D 5, P 15, V 10, A 86, C 21.

8. *Silurus Garua* pinna caudali bifida; cirrhis 8; radiis pinnæ dorsalis 8, analis 29.

B 8, D $\frac{1}{8}$, P $\frac{1}{12}$, V 6, A 29, C 17+.

Genus V. Malapterurus cute lævi, glutinosa, alepidota; ore cirrhoso; pinna dorsi unica, postica, brevi, adiposa.

Species 1. *Malapterurus Coila* cirrhis 8.

B—, D 0, P 16, V 7, A 71, C 16+.

Genus VI. Ageneiosus cute lævi, glutinosa, alepidota; ore imberbi; pinnis dorsi duabus, posteriore adiposa.

Species 1. *Ageneiosus Mino* radiis pinnæ dorsalis 8, analis 21.

B 5, D $\frac{2}{8}$ et 0, P $\frac{1}{9}$, V 6, A 21, C 19+.

Genus VII. Pimelodus cute glutinosa, alepidota; ore cirrhoso; pinnis dorsi duabus, posteriore adiposa.

Species 1. *Pimelodus Silondia* cirrhis duobus; corpore opaco.

B 10, D $\frac{1}{8}$ et 0, P $\frac{1}{14}$, V 6, A 41, C 17+.

2. *Pimelodus Chandramara* cirrhis 2; corpore diaphano.

B—, D $\frac{1}{8}$ et 0, P $\frac{1}{-}$, V 6, A 17, C —.

Species 3. *Pimelodus Pangasius* pinna caudali bifida; cirrhis 4; aculeis dorsi pectorisque curtis.

B 10, D $\frac{2}{9}$ et 0, P $\frac{1}{12}$, V 6, A 31, C 17+.

4. *Pimelodus Rita* pinna caudali biloba; cirrhis 6 capite brevioribus; radiis dorsi 8, ani 12; aculeo dorsi integerrimo; corpore opaco, emaculato.

B—, D $\frac{2}{8}$ et 0, P $\frac{1}{12}$, V 8, A 12, C 15+.

5. *Pimelodus Gagora* pinna caudæ bifida; cirrhis 6 capite brevioribus; radiis dorsi 9, ani 18; aculeo dorsi curto, utrinque serrato; lateribus opacis, emaculatis.

B 4, D $\frac{2}{9}$ et 0, P $\frac{1}{12}$, V 6, A 18, C 18+.

6. *Pimelodus Sagor* pinna caudali bifida; cirrhis 6, pisce demidio brevioribus; radiis dorsi 9, ani 18; aculeo dorsi utrinque serrato, curto; lateribus opacis, supra lineas laterales solitarias fasciatis.

B 4, D $\frac{2}{9}$ et 0, P $\frac{1}{11}$, V 6, A 18, C 16+.

7. *Pimelodus Arius* pinna caudali bifida; cirrhis 6 capite non longioribus; radiis dorsi 8, ani 22; aculeo dorsi longius ultra pinnæ membranam producto, utrinque serrato; palato dentato; lateribus emaculatis, opacis.

B 5, D $\frac{2}{8}$ et 0, P $\frac{1}{11}$, V 6, A 22, C 15+.

8. *Pimelodus Jatius* pinna caudæ bifida; cirrhis 6 capite non longioribus; radiis dorsi 8, ani 21; aculeo dorsi longius ultra pinnæ membranam producto, utrinque serrato; palato edentulo; lateribus opacis.

B 5, D $\frac{2}{8}$ et 0, P $\frac{1}{11}$, V 6, A 21, C 15+.

9. *Pimelodus Nenga* pinna caudæ bifida; cirrhis 6 capite longioribus; radiis dorsii 9, ani 20; aculeo dorsi longius ultra pinnæ membranam producto, utrinque serrato; palato dentato; corpore opaco, lineis brevibus fasciato.

B 4, D $\frac{2}{9}$ et 0, P $\frac{1}{12}$, V 6, A 20, C 18+.

10. *Pimelodus Sona* pinna caudæ bifida; cirrhis 6 capite longioribus; radiis dorsi 9, ani 18; aculeo dorsi longius ultra

pinnæ membranam producto, utrinque serrato; corpore opaco, emaculato.

B 4, D $\frac{2}{9}$ et 0, P $\frac{1}{12}$, V 6, A 18, C 18+.

Species 11. *Pimelodus viridescens* pinna caudæ bifida; cirrhis 6; radiis dorsi 7, ani 13; aculeo dorsi integerrimo; dorsi rubicundi fasciis 5 viridescentibus; lateribus opacis, emaculatis.

B 6, D $\frac{2}{7}$ et 0, P $\frac{1}{9}$? V 6, A 13, C 20+.

12. *Pimelodus Rama* pinna caudæ bifida; cirrhis 6 capite brevioribus; corpore diaphano.

B—, D $\frac{4}{8}$ et 0, P $\frac{1}{-}$, V 6, A 15, C—.

13. *Pimelodus Tengana* pinna caudæ bifida; cirrhis 8 capite brevioribus; corpore diaphano, supra punctato; radiis dorsi 8, ani 14.

B 6, D $\frac{2}{8}$ et 0, P $\frac{1}{-}$, V 6, A 14, C—.

14. *Pimelodus Urua* pinna caudæ bifida; cirrhis 8 capite longioribus; corpore diaphano, supra punctato; radiis dorsi 5, ani 43.

B 9? D $\frac{1}{5 \text{ seu } 6}$ et 0, P $\frac{1}{10}$? V 7? A 43? C 24?

15. *Pimelodus Batasio* pinna caudæ bifida; cirrhis 8 capite brevioribus; corpore diaphano, vittis utrinque 2 nigro punctatis; radiis dorsi 9, ani 16.

B 6? D $\frac{2}{9}$ et 0, P $\frac{1}{-}$, V 6, A 16, C 20?

16. *Pimelodus Angius* pinna caudæ bifida; cirrhis 8 capite longioribus; corpore subdiaphano, vittis punctatis utrinque 3; radiis dorsi 7, ani 42.

B—, D $\frac{1}{7}$ et 0, P $\frac{1}{7}$, V 6, A 42? C—.

17. *Pimelodus Carcio* pinna caudæ bifida; cirrhis 8, duobus tantum capite longioribus; corpore subopaco, vittis punctatis utrinque tribus; radiis dorsi 8, ani 14.

B 6? D $\frac{2}{8}$ et 0, P $\frac{1}{7}$? V 6, A 14, C—.

18. *Pimelodus Tengara* pinna caudæ bifida; cirrhis 8, quorum 4 capite longiores; corpore opaco, lævi, vittis punctatis utrinque 4; radiis dorsi 8, ani 10.

B 6? D $\frac{1}{8}$ et 0, P $\frac{1}{8}$, V 6, A 10, C—.

Species 19. *Pimelodus Telchita* pinna caudæ bifida; cirrhis 8 capite brevioribus; corpore opaco, aspero, rubicundo, vitta utrinque obsoleta; radiis dorsi 8, ani 13.

B 4 seu 5, D $\frac{2}{8}$ et 0, P $\frac{1}{10}$, V 6, A 13, C 19+.

20. *Pimelodus Bagarius* pinna caudæ biloba; cirrhis 8 capite non longioribus; corpore opaco, scabro, maculis nigris nebuloso; radiis dorsi 8, ani 13.

B 11, D $\frac{2}{8}$ et 0, P $\frac{1}{15}$, V 6, A 13, C 17+.

21. *Pimelodus Cavia* pinna caudæ biloba, lobo inferiore longiore; cirrhis 8 capite non longioribus; corpore opaco, scabro, nebuloso; radiis dorsi 5, ani 9.

B 6? D $\frac{1}{5}$ et 0, P $\frac{1}{7}$? V 6, A 9, C 13?

22. *Pimelodus Hara* pinna caudæ lunata; cirrhis 8, quorum 2 capite paulo longiores; corpore opaco, scabro, fusco, nebuloso; maculis pinnarum pellucidis; radiis dorsi 7, ani 10.

B—, D $\frac{1}{7}$ et 0, P $\frac{1}{8}$, V 6, A 10, C—.

23. *Pimelodus Conta* pinna caudæ biloba, lobo superiore subulato; cirrhis 8 capite brevioribus; corpore opaco, scabro, fusco; maculis pinnarum pellucidis; radiis dorsi 5, ani 10.

B—, D $\frac{1}{5}$ et 0, P $\frac{1}{8}$, V 6, A 10, C 16?

24. *Pimelodus Botius* pinna caudæ bifida; cirrhis 8 capite brevioribus; corpore opaco, fuscescente, scabro; radiis dorsi 6, ani 11.

B—, D $\frac{1}{6}$ et 0, P $\frac{1}{8}$? V 5, A 11, C 18+.

25. *Pimelodus Nangra* pinna caudæ bifida; cirrhis 8, quorum 2 ad anum pertingunt; corpore subopaco, lævi, emaculato, evittato; radiis dorsi 8, ani 9.

B 6? D $\frac{2}{8}$ et 0, P $\frac{1}{9}$? V 6, A 9, C 16+.

26. *Pimelodus Murius* pinna caudæ bifida; cirrhis 8, capite non longioribus; corpore opaco, lævi, supra viridi; radiis dorsi 8, ani 42.

B 5? D $\frac{1}{8}$ et 0, P $\frac{1}{11}$, V 6, A 42, C—.

27. *Pimelodus Vacha* pinna caudæ bifida; cirrhis 8 caput lon-

gitudine æquantibus; corpore subopaco, lævi, supra viridi; radiis dorsi 7, ani 50.

B 10, D $\frac{1}{7}$ et 0, P $\frac{1}{16}$, V 6, A 50, C 17+.

Species 28. *Pimelodus Gagata* pinna caudæ bifida; cirrhis 8, quorum 2 capite paulo longiores; corpore opaco, lævi, flavescente; radiis dorsi 8, ani 17.

B 5, D $\frac{2}{8}$ et 0, P $\frac{1}{10}$, V 6, A 17, C 18+.

29. *Pimelodus Mangois* pinna caudæ biloba; cirrhis 8 capite paulo longioribus; corpore opaco, lævi, subolivaceo; radiis dorsi 7, ani 8.

B 8? D $\frac{1}{7}$ et 0, P $\frac{1}{8}$, V 6, A 8, C 26.

30. *Pimelodus Gulio* pinna caudæ subbifida; cirrhis 8, quorum 2 ad finem pinnæ dorsalis anterioris, 4 ultra caput pertingunt; corpore opaco, lævi, supra fuscescente; radiis dorsi 8, ani 15.

B—, D $\frac{2}{8}$ et 0, P $\frac{1}{9}$, V 6, A 15, C 18+.

31. *Pimelodus Menoda* pinna caudæ subbifida; cirrhis 8, quorum 2 ad finem pinnæ dorsalis anterioris, 4 ultra caput pertingunt; corpore opaco, lævi, supra fuscescente; radiis dorsi 9, ani 13.

B—, D $\frac{2}{9}$ et 0, P $\frac{1}{9}$, V 6, A 13, C 18+.

32. *Pimelodus Cavasius* pinna caudæ biloba, acuta; cirrhis 8, quorum 2 cauda longiores; corpore subopaco, lævi, supra fuscescente; radiis dorsi 8, ani 13.

B 7? D $\frac{1}{8}$ et 0, P $\frac{1}{11}$? V 6, A 13, C—.

33. *Pimelodus Aor* pinna caudæ biloba, supra falcata; cirrhis 8, quorum 2 pinnam caudæ, 2 nucham pertingunt; corpore subopaco, lævi, supra fuscescente; radiis dorsi 9, ani 12.

B 14, D $\frac{2}{9}$ et 0, P $\frac{1}{11}$, V 6, A 12, C 17+.

Genus IX. Sisor corpore glutinoso alepidoto; cauda nuda; ore cirrhoso; pinnis dorsi 2, utraque radiata.

Species 1. *Sisor Rabdophorus.*

B 4? D $\frac{1}{7}$ et $\frac{1}{1}$, P $\frac{1}{12}$, V 7, A 6, C 10.

Genus X. Osmerus corpore squamoso; pinnis dorsi geminis, posteriore adiposa; ore imberbi; dentibus maxillarum vallidis.

Species 1. *Osmerus? Neherius* radiis dorsi 12, ani 15; pinna caudæ trifida.

B 0? D 12 et 0, P 12, V 9, A 15, C 18?

Genus XI. Esox ore magno imberbi; pinna dorsi unica, lumbari, radiata; dentibus maxillarum vallidis; ventre obtuso.

Species 1. *Esox Panchax* pinna caudæ integra; maxillis equalibus, rotundatis; radiis dorsi 6, ani 14.

B 4, D 6, P 16, V 6, A 14, C 16.

2. *Esox Ectuntio* pinna caudæ integra; maxillis in rostrum elongatis, subæqualibus; corpore capiteque alepidotis; radiis dorsi 13, ani 11.

B 6, D 13, P 9, V 6, A 11, C 15+.

3. *Esox Cancila* pinna caudæ integra; maxillis in rostrum elongatis, subæqualibus; corpore capiteque alepidotis, radiis dorsi 17, ani 18.

B 13, D 17, P 9, V 6, A 17 seu 18, C 15+.

Genus XII. Mugil corpore squamoso; pinnis dorsalibus radiatis duabus; digitis ante pinnas pectorales liberis nullis; maxilla inferiore carinata.

Species 1. *Mugil Parsia* radiis pinnæ dorsalis posterioris 9; aculeis 3, radiisque mollibus 8 in pinna anali; caudæ pinna lunata; naribus geminatis; operculis branchiostegis arcuatis; capite declive; lateribus glabris, non vittatis.

B 6, D $\frac{4}{4}$ et 9, P 14, V 6, A $\frac{3}{11}$, C 14+.

2. *Mugil Cascasia* radiis pinnæ dorsalis posterioris 8; aculeis 2, radiisque mollibus 8 in pinna anali; pinna caudæ lunata; naribus geminis; operculis branchiostegis arcuatis; capite declive; lateribus scabris non vittatis.

B 3? D $\frac{4}{4}$ et 8, P —, V $\frac{1}{6}$, A $\frac{2}{10}$, C—.

3. *Mugil Albula?* radiis pinnæ dorsalis posterioris 9; aculeis 4, radiisque mollibus 9 in pinna anali; pinna caudæ lunata; naribus geminis; operculis branchiostegis arcuatis; capite declive; lateribus glabris, argenteo versicolore vittatis.

B 6, D $\frac{4}{4}$ et 9, P 15, V $\frac{1}{6}$, A $\frac{4}{13}$, C 14+.

Species 4. *Mugil Cephalus?* lateribus nigro vittatis; oculis planis; maxillis edentulis, inferiore breviore; squamis integerrimis.

B—, D $\frac{4}{4}$ et 9, P 15, V 6, A 12, C 14+.

5. *Mugil Corsula* radiis pinnæ dorsalis posterioris 9; aculeis 3, radiisque mollibus 9 in pinna anali; pinna caudæ lunata; capite adscendente; oculis protuberantibus; laterum vittis e punctis nigris conflatis.

B 4, D $\frac{4}{4}$ et $\frac{1}{9}$, P 15, V 6, A $\frac{3}{12}$, C 14+.

Genus XIII. Atherina pinnis dorsalibus duabus radiatis; digitis ante pinnas pectorales liberis nullis; radiis in pinnis ventralibus et in membranis branchiostegis singulis minus 8; palato edentulo; corpore elongato, subdiaphano, vitta utrinque argentea.

Species 1. *Atherina Danius* radiis pinnæ dorsalis anterioris circiter 4, posterioris 8, analibus 10.

B—, D 4? et 8? P 7, V 6, A 10? C—.

Genus XIV. Polynemus pinnis dorsalibus radiatis duabus; digitis ante pinnas pectorales nonnullis liberis.

Species 1. *Polynemus Teria* digitis 4; pinna caudali biloba.

B 7, D $\frac{8}{8}$ et $\frac{1}{15}$, P 17, V $\frac{1}{6}$, A $\frac{2}{17}$, C 15+.

2. *Polynemus Sele* digitis 5 ad anum pertingentibus; lateribus non vittatis; linea laterali per lobum pinnæ caudalis inferius deflexa.

B 5, D $\frac{7}{7}$ et 14, P 13, V $\frac{1}{6}$, A 13, C 20+.

3. *Polynemus Risua* digitis 7, secundo longissimo, corporis longitudinem duplo superante; aculeis dorsalibus 7; radiis in pinna dorsi posteriore 17, in pinna ani 15; dorso viridi; ventre argenteo.

B 7, D $\frac{7}{7}$ et 17, V $\frac{1}{6}$, A 15, C 17+.

4. *Polynemus Toposui* digitis 7, priore longissimo, corporis longitudinem duplo superante; aculeis dorsalibus 7; radiis in pinna dorsi posteriore 16, in pinna ani 16; dorso viridi, ventre argenteo.

B 7, D $\frac{7}{7}$ et 16, P 16, V $\frac{1}{6}$, A 16, C—.

5. *Polynemus aureus* digitis 7, longissimo corporis longitudinem

duplo superante; aculeis dorsalibus 7; radiis in pinna dorsi posteriore 15, in pinna ani 14; colore aureo, supra fuscescente.

B 7, D $\frac{7}{7}$ et 15, P 17, V $\frac{1}{6}$, A 14, C 17+.

Genus XV. MYSTUS radiis branchiostegis utrinque plus tribus; ventre carinato; pinna ani longa cum caudali unita; dorso monopterygio.

Species 1. *Mystus Ramcarate* maxilla superiore inermi; radiis pinnarum pectoralium senis primariis setaceis, membrana multoties longioribus; radiis pinnarum ventralium denis.

B 10, D 14, P 12, V $\frac{1}{10}$, A et C 111?

2. *Mystus Kapirat* maxilla superiore inermi; radiis pinnarum ventralium unitarum geminis.

B 4? D 9, P 12, V 1, A et C 110?

3. *Mystus Chitala* maxilla superiore inermi; radiis pinnarum ventralium subdistinctarum utrinque 4.

B—, D 9, P 14? V 4, A et C 120–130.

Genus XVI. CLUPEA maxillis dentatis; pinna dorsali radiata, unica; corpore oblongo, squamoso; ventre carinato; pinnis ani caudæque distinctis.

Species 1. *Clupea Purava* mystacibus elongatis, ensiformibus; radiis analibus 46; pinna caudali bifida; corpore immaculato.

B 11, D 13, P 13, V 7, A 46, C 18+.

2. *Clupea Phasa* radio pinnarum pectoralium primario longissimo setaceo; lobis pinnæ caudalis utrisque acutis; radiis dorsalibus 14, analibus 78 circiter.

B 3? D 14, P 13, V 7, A 75–78, C—.

3. *Clupea Telara* radio pinnarum pectoralium primario longissimo setaceo; lobo pinnæ caudalis superiore truncato.

B 13, D 14, P 15, V 7, A 74, C 19+.

Genus XVII. CLUPANODON maxillis edentulis; corpore oblongo, squamoso, supra obtuso, subtus carinato; pinna dorsi unica brevi; pinna ani ab caudali distincta.

Species 1. *Clupanodon Ilisha* radiis dorsalibus 20, analibus 21, ventralibus 8.

B 6, D 20, P 15, V 8, A 21, C 19+.

Species 2. *Clupanodon Chacunda* radiis dorsalibus 18, analibus 19, ventralibus 8, branchiostegis 5; maxilla inferiore breviore.

B 5, D 18, P 15, V 8, A 19, C 19+.

3. *Clupanodon Manmina* radiis dorsalibus 14, analibus 24; macula in humero utroque nigra.

B 5, D 14, P 15, V 8, A 24, C 19+.

4. *Clupanodon Chapra* radiis dorsalibus 17, analibus 24; macula in humero utroque nigra.

5. *Clupanodon Cortius* radiis dorsalibus 15, analibus 24; humeris immaculatis.

6. *Clupanodon Chanpole* radiis dorsalibus 15, analibus 21; macularum nigrarum serie in humero utroque.

B 6, D 15, P 13, V 8, A 21, C 19+.

7. *Clupanodon Cagius* radiis dorsalibus 15, analibus 23; serie macularum nigrarum in humero utroque; dorso carinato.

B 6, D 15, P 14? V 8, A 23, C—.

8. *Clupanodon Motius* radiis dorsalibus 15, analibus 43; maxilla superiore integerrima.

B 5? D 15, P 14 vel 15, V 7, A 43? C 18+.

Genus XVIII. Corica dentibus nullis; corpore oblongo, subtus carinato, serrato; pinna dorsali solitaria, brevi; pinnis analibus duabus.

Species 1. *Corica soborna.*

B 5? D 13, P 12, V 7, A 13 et 4, C 18?

Genus XIX. Cyprinodon corpore oblongo, squamoso, compressissimo; ventre obtuso; pinna dorsali solitaria, submedia; dentibus in maxilla utraque.

Species 1. *Cyprinodon Cundinga* radio pinnæ dorsalis postremo longissimo.

B 25, D 19, P 15, V 10, A 26, C 19+.

Genus XX. Cyprinus pinna dorsali solitaria, radiata; radiis dorsalibus pectoralibusque muticis; maxillis edentulis; radiis branchiostegis ternis; cirrhis vel nullis vel duobus, vel quatuor.

N. B. In unica specie radios branchiostegos 4, in altera duos notavi.

Subgenus I. Chela Cyprini Clupanodontibus affines ventre inter pinnas ventrales et analem carinato.

Species 1. *Cyprinus Cachius*, Chela radiis pinnarum ventralium ternis, analis 26 circiter, dorsalis 7 circiter.

B 3? D 7, P 8, V 3, A 26, C—.

2. *Cyprinus Atpar*, Chela radiis pinnarum ventralium quinis, analis 25, dorsalis 9.

B 3, D 9, P 11 vel 12, V 5, A 25, C—.

3. *Cyprinus Laubuca*, Chela radiis pinnarum ventralium septenis; analis 24, dorsalis 10.

B 3, D 10, P 12, V 7, A 24, C—.

4. *Cyprinus Phulo*, Chela radiis pinnarum ventralium octonis, analis 20, dorsalis 9.

B 3, D 9, P 11, V 8, A 20, C 18+.

5. *Cyprinus Gora*, Chela radiis pinnarum ventralium octonis, analis 17, dorsalis 9; linea laterali immaculata.

B 3, D 9, P 13? V 8, A 17, C —.

6. *Cyprinus Morar*, Chela radiis pinnarum ventralium octonis, analis 12, dorsalis 10; ventris carina integerrima.

B 3, D 10, P 15, V 8, A 12, C 20+.

7. *Cyprinus Bacaila*, Chela radiis pinnarum ventralium novenis, analis 16, dorsalis 9; pinnis hyalinis.

B 3, D 9, B 12, V 9, A 16, C 19+.

Subgenus II. Barilius, Cyprini Truttis affines corpore compressissimo, fasciis demidiatis, vel maculis pluribus vagis vario; pinna dorsali lumbari; linea laterali decurva, ventri parallela.

8. *Cyprinus Barila*, Barilius imberbis lateribus semifasciatis; radiis in pinna dorsi 8, in anali 13.

B 3, D 8, P 14, V 8, A 13, C—.

9. *Cyprinus Chedrio*, Barilius imberbis lateribus semifasciatis; radiis in pinna dorsi 9, in anali 13.

B 3, D 9, P 14, V 8, A 13, C—.

10. *Cyprinus Barna*, Barilius imberbis lateribus semifasciatis; radiis in pinna dorsi 9, in anali 12; linea laterali superiore auro vittata.

B 3, D 9, P 13? V 9, A 12, C 19+.

Species 11. *Cyprinus Vagra*, Barilius imberbis lateribus semifasciatis; radiis in pinna dorsali 9, in anali 14.

B—, D 9, P 13, V 9, A 14, C 16+.

Cyprinus Bendilisis, Barilius cirrhis 2; lateribus semifasciatis; radiis in pinna dorsali 9, in anali 11.

B—, D 9, P 13, V 9, A 11, C 19+.

12. *Cyprinus Phacra*, Barilius cirrhis 4; lateribus semifasciatis; radiis in pinna dorsi 9, in anali 10.

B—, D 9, P 13, V 9, A 10, C 20+.

13. *Cyprinus Cocsa*, Barilius cirrhis 4; lateribus et semifasciatis, et medio squamarum singularum nigro maculatis; radiis in pinna dorsali 9, in anali 11.

B 3, D 9, P 13? V 9, A 11, C 19.

14. *Cyprinus Chedra*, Barilius imberbis macula in medio squamarum in lateribus singularum; radiis in pinna dorsali 10, in anali 11.

B 3, D 10, P 14? V 9, A 11, C 18+.

15. *Cyprinus Tila*, Barilius imberbis macula in medio squamarum in lateribus singularum; radiis in pinna dorsali 8, in anali 10.

B 3, D 8, P 14 seu 15, V 9, A 10, C 18+.

16. *Cyprinus Bola*, Barilius imberbis maculis pluribus diffusis, vittaque aurea in latere utroque; radiis in pinna dorsali 10, in anali 11.

B 3, D 10, P 16? V 9, A 11, C 19+.

17. *Cyprinus Goha*, Barilius imberbis maculis pluribus diffusis sparsis in latere utroque; radiis in pinna dorsali 10, in anali 13.

B 3, D 10, P—, V 9, A 13, C—.

Subgenus III. BANGANA Cyprini Mugilibus affines maxilla inferiore ntus carinata.

18. *Cyprinus Dero*, Bangana cirrhis duobus, radiis in pinna dorsali 13, in anali 7.

B 3, D 13, P 18? V 9, A 7, C 19+.

Species 19. *Cyprinus Mrigala*, Bangana cirrhis duobus; radiis in pinna dorsali 16, in anali 8.

B 3, D 16, P 17, V 9, A 8, C 19+.

20. *Cyprinus Reba*, Bangana cirrhis 2; radiis in pinna dorsali 11, in anali 8; labiis carnosis.

B 3, D 11, P 17, V 9, A 8, C 19+.

21. *Cyprinus Elanga*, Bangana cirrhis vix ullis; radiis in pinna dorsali 9, in anali 8; labiis obseletis.

B 3, D 9, P 16, V 9, A 8, C 19+.

22. *Cyprinus Bata*, Bangana imberbis radiis in pinna dorsali 12, in anali 8; labio inferiore integerrimo, erecto; lobo pinnæ caudalis superiore longiore; pinnis omnibus pallidis.

B 3, D 12, P 17, V 9, A 8, C 19+.

23. *Cyprinus Acra*, Bangana imberbis radiis in pinna dorsali 11, in anali 8; labio inferiore integerrimo, erecto; lobo pinnæ caudalis superiore longiore; pinnis inferioribus rubicundis.

B 3, D 11, P 18 vel 19, V 9, A 8, C 19+.

24. *Cyprinus Cura*, Bangana imberbis radiis in pinna dorsali 12, in anali 7; labio inferiore integerrimo, erecto; lobis pinnæ caudalis æqualibus.

B 3, D 12, P—, V 9, A 7, C—.

25. *Cyprinus Pangusia*, Bangana imberbis radiis in pinna dorsali 14, in anali 8; labio inferiore reflexo, integerrimo; rostro poris punctato.

B 3, D 14, P 17, V 9, A 8, C 19+.

26. *Cyprinus Ariza*, Bangana imberbis radiis in pinna dorsali 12, in anali 7; labio inferiore reflexo, integerrimo; rostro lævi.

B 3, D 12, P 16, V 9, A 7, C—.

27. *Cyprinus Boga*, Bangana imberbis radiis in pinna dorsali 12, in anali 8; labio inferiore crenato; rostro poris punctato.

B 3, D 12, P—, V 9, A 8, C—.

Subgenus IV. CYPRINI VERI absque ulla ad aliud genus affinitate; corpore magno, crasso, macularum insignium experte; linea laterali submedia.

Species 28. *Cyprinus Catla*, Cyprinus verus imberbis radiis in pinna dorsali 18, in anali 8; colore subargenteo, dorso fusco-viridescente; labiis integerrimis.

B 3, D 18, P—, V 9, A 8, C 19+.

29. *Cyprinus Curchius*, Cyprinus verus imberbis radiis in pinna dorsali 15, in anali 7; squamis minutis.

B 3, D 15, P 18, V 9, A 7, C 19+.

30. *Cyprinus Cursa*, Cyprinus verus cirrhis 2; radiis in pinna dorsali 16, in anali 8; squamis parvis; rostro poris callosis punctato.

B 3, D 16, P 17, V 9, A 8, C 91+.

31. *Cyprinus Cursis*, Cyprinus verus cirrhis 4; radiis in pinna dorsali 16, in anali 8; squamis parvis; rostro poris callosis punctato; labiis fimbriatis.

B 3, D 16, P 17, V 9, A 8, C 19+.

32. *Cyprinus Gonius*, Cyprinus verus cirrhis 4; radiis in pinna dorsali 15, in anali 7; squamis mediocribus; rostro poris callosis punctato; labiis fimbriatis; lateribus subvittatis.

B 3, D 15, P 17, V 9, A 7, C—.

— *Cyprinus Curmuca*, Cyprinus verus cirrhis 2; radiis in pinna dorsali 11, in anali 8; squamis mediocribus; capite anterius tuberculato; radio dorsali tertio integerrimo.

B 3, D 11, P 16, V 9, A 8, C 18+.

33. *Cyprinus Chagunio*, Cyprinus verus cirrhis 4; radiis in pinna dorsali 12, in anali 8; squamis magnis; capite anterius tuberculato; radio dorsali tertio pone serrato.

B 3, D 12, P 17, V 10, A 8, C 19+.

34. *Cyprinus Calbasu*, Cyprinus verus cirrhis 4; radiis in pinna dorsali 16, in anali 8; squamis magnis; rostro lævi, imperforato; labiis serratis.

B 3, D 16, P 18, V 9, A 8, C 19+.

35. *Cyprinus Nancar*, Cyprinus verus cirrhis 4; radiis in pin-

na dorsali 20, in anali 8; squamis magnis equalibus; rostro lævi, imperforato; labiis integerrimis.

B 3, D 20, P 18? V 9, A 8, C 20+.

Species 36. *Cyprinus Nandina*, Cyprinus verus cirrhis 4; radiis in pinna dorsali 26, in anali 8; squamis magnis; rostro lævi; labiis fimbriatis.

B 3, D 26 vel in aliis 23, P 16, V 9, A 8, C 23.

37. *Cyprinus Rohita*, Cyprinus verus cirrhis 4; radiis in pinna dorsali 15, in anali 7; squamis magnis; rostro lævi imperforato; labiis serratis, inferiore reflexo.

B 3, D 15, P 18, V 10, A 7, C 19+.

38. *Cyprinus Putitora*, Cyprinus verus cirrhis 4; radiis in pinna dorsali 11, duobus prioribus simplicibus, in pinna anali 7; squamis maximis; rostro lævi imperforato; labiis integerrimis; pinnis pallidis.

B 3, D 11, P 15, V 9, A 7, C 19+.

Varietas cirrhis 2; radiis in pinna dorsali 10, in anali 8.

39. *Cyprinus Tor*, Cyprinus verus cirrhis 4; radiis in pinna dorsi 11, quorum 3 priores simplices, in anali 8; squamis maximis; rostro lævi imperforato; labiis integerrimis; pinnis inferioribus rubicundis.

B 3, D 11, P 18, V 9, A 8, C 19+.

40. *Cyprinus Mosal*, Cyprinus verus cirrhis 4; radiis in pinna dorsali 13, quorum 4 priores simplices, in anali 8; squamis maximis; rostro lævi imperforato; labiis integerrimis; pinnis pallidis.

B 3, D 13, P 17? V 9, A 8, C 19+.

41. *Cyprinus Sarana*, Cyprinus verus cirrhis 4; radiis in pinna dorsali 10, longiore simplicium pone serrato, in anali 8; squamis magnis; maxillis subæqualibus; labio inferiore erecto, integerrimo; pinnis pallidis.

B 3, D 10, P 16, V 9, A 8, C 19+.

Subgenus V. PUNTIUS Cyprini absque ulla ad aliud genus affinitate; corpore parvo, maculis paucis magnis saturatis insigni.

Species 42. *Cyprinus Sophore,* Puntius corpore opaco; macula diffusa aurea in operculis utrinque, nigra in caudæ extremitate utrinque, et ad pinnæ dorsalis insertionem; radio pinnæ dorsalis secundo pone integerrimo.

B 3, D 10, P 14, V 9, A 7, C 19+.

43. *Cyprinus Chola,* Puntius corpore opaco; macula aurea diffusa in operculis utrinque, nigra prope caudæ finem; radio pinnæ dorsalis secundo pone integerrimo.

B 3, D 10, P 13? V 9, A 7, C 19+.

44. *Cyprinus Terio,* Puntius corpore opaco; macula aurea diffusa in operculis utrinque, nigra in medio caudæ; radiis pinnæ dorsalis 3 primariis indivisis integerrimisque.

B 3, D 11, P 13? V 9, A 8, C 19+.

45. *Cyprinus Ticto,* Puntius corpore opaco; maculis utrinque nigris duabus, una supra pinnam pectoralem, altera caudæ versus finem; pinnæ dorsalis maculatæ radio secundo pone serrato.

B 3, D 10, P 12, V 9, A 7, C 20+.

46. *Cyprinus Titius,* Puntius corpore opaco; maculis nigris utrinque 2 prope lineam lateralem; pinnæ dorsalis immaculatæ radii secundo pone integerrimo.

47. *Cyprinus Joalius,* Puntius corpore opaco; macula nigra lunata versus finem caudæ utrinque; radio pinnæ dorsalis secundo pone integerrimo; labio inferiore serrato.

B 3, D 12, P—, V 9, A 8, C—.

48. *Cyprinus Pausio,* Puntius corpore opaco; macula nigra versus caudæ finem utrinque; radio pinnæ dorsalis secundo, labioque inferiore integerrimis.

B 3, D 13, P—, V 9, A 7, C—.

49. *Cyprinus Conchonius,* Puntius corpore opaco; macula utrinque nigra in caudæ medio; pinnæ dorsalis immaculatæ radio secundo pone serrato.

B—, D 10, P 10? V 9, A 8, C 19+.

50. *Cyprinus Puntio,* Puntius corpore subopaco; annulo circa

finem caudæ nigro; macula nigra ad insertiones pinnarum dorsi anique.

B—, D 9, P 12, V 8, A 6, C—.

Species 51. *Cyprinus Phutunio,* Puntius corpore subdiaphano; maculis in utroque latere quinque nigris; radio pinnæ dorsalis secundo pone serrato.

B—, D 10, P 9? V 8, A 7, C—.

52. *Cyprinus Gelius,* Puntius corpore diaphano flavescente; operculis peritonæo spinaque dorsali argenteis; maculis utrinque magnis nonnullis; radio pinnæ dorsalis 2do pone serrato.

B 3, D 10, P—, V 9, A 8, C—.

53. *Cyprinus Canius,* Puntius corpore diaphano rubescente; operculis peritonæo spinaque dorsali argenteis; maculis utrinque magnis nigris nonnullis; radio pinnæ dorsalis secundo pone serrato.

B 3, D 10, P 8? V 9, A 7, C—.

Subgenus VI. Danio Cyprini absque ulla ad aliud genus affinitate; corpore parvo, vittato, immaculato.

54. *Cyprinus Dangila,* Danio vittis cæruleis anastomozantibus; corpore compresso; cirrhis 4, quorum 2 capite longiores.

B 3, D 13, P 12? V 7, A 17, C 28.

55. *Cyprinus Rerio,* Danio vittis utrinque cæruleis argenteisque nonnullis; corpore compressissimo; cirrhis 4, quorum 2 capite paulo longiores.

B 3, D 8, P 9? V 7, A 16, C—?

56. *Cyprinus Chapalio,* Danio vittis cæruleis utrinque 3; corpore compressiusculo; cirrhis 2 capite brevioribus.

B 3, D 9, P 12, V 7, A 13, C 18+.

57. *Cyprinus Danrica,* Danio vitta infra lineam lateralem utrinque punctata, nigrescente; corpore compressiusculo; cirrhis 4, quorum duo longissimi; radiis pinnarum ventralium primariis setaceo elongatis.

B 3, D 8, P 13, V 8, A 8, C 19+.

Species 58. *Cyprinus Jogia*, Danio vitta infra lineam lateralem utrinque punctata, nigrescente; corpore compressissimo; cirrhis 4, quorum 2 longissimi; radiis pinnarum ventralium primariis curtis.

B 3, D 8, P 10? V 7, A 7, C 19+.

59. *Cyprinus Sutiha*, Danio vitta per lineam lateralem utrinque argentea; corpore compresso; cirrhis 4, quorum duo longissimi; radiis pinnarum pectoralium ventraliumque primariis setaceo productis.

B 3, D 3, P 10, V 7, A 7, C 19+.

60. *Cyprinus Daniconius*, Danio vitta infra lineam lateralem utrinque punctis nigrescente; corpore compressiusculo; capite imberbi; radiis pinnæ dorsalis 9, analis 7; linea laterali recta.

B 3, D 9, P 13, V 9, A 7, C 19+.

61. *Cyprinus Anjana*, Danio vittis nigro punctatis utrinque duabus; corpore compressiusculo; capite imberbi; radiis in pinnæ dorsali 9, anali 7.

B 3, D 9, P 13? V 9, A 7, C 19+?

62. *Cyprinus Rasbora*, Danio vitta punctata utrinque unica; capite imberbi; corpore compressiusculo; radiis pinnæ dorsalis 9, analis 8; linea laterali ventri parallela.

B—, D 9, P 13, V 9, A 8, C 19+.

63. *Cyprinus Angra*, Danio lateribus flavescentibus; vitta unica utrinque nigro punctato; cirrhis minutis duobus; corpore compresso.

B 3, D 12, P 16? V 9, A 8, C 19+.

Subgenus VII. Marulius Cyprini absque ulla ad aliud genus affinitate; corpore parvo; dorsi ventrisque colorum commissura invicem serrata.

64. *Cyprinus Morala*, Morulius cirrhis 4; radiis in pinna dorsali 13, in anali 8.

B 3, D 13, P 16? V 9, A 8, C 19+.

Species 65. *Cyprinus Pausius*, Morulius cirrhis 4; radiis in pinna dorsali 14, in anali 7.

B 3, D 14, P 16? V 9, A 7, C 19+.

66. *Cyprinus Musiha*, Morulius imberbis; radiis in pinna dorsali 13, in anali 8; corpore compresso.

B 3, D 13, P 16? V 9, A 8, C 19+.

Subgenus VIII. Cabdio Cyprini absque ulla ad aliud genus affinitate; corpore parvo, compresso; absque maculis, vittis, vel alia colorum nota insigni.

67. *Cyprinus Jaya*, Cabdio radiis in pinna dorsali 8, in anali 9, in pectoralibus 13 circiter, in ventralibus 8; lineis utrinque lateralibus 2.

B 3, D 8, P 13? V 8, A 9, C 17+.

68. *Cyprinus Mola*, Cabdio radiis in pinna dorsali 9, in anali 7, in pectoralibus 15, in ventralibus 9; lineis lateralibus 2 utrinque.

B 3, D 9, P 15, V 9, A 7, C 19+.

69. *Cyprinus Hoalius*, Cabdio radiis in pinna dorsali 9, in anali 10, in ventralibus 9; linea laterali solitaria; ventre argenteo.

B—, D 9, P—, V 9, A 10, C—.

70. *Cyprinus Borelio*, Cabdio radiis in pinna dorsali 9, in anali 11, in ventralibus 8; linea laterali gemina; ventre flavo.

B 3, D 9, P 13? V 8, A 11, C 18+.

71. *Cyprinus Solio*, Cabdio radiis in pinna dorsali 9, in anali 11, in ventralibus 8; linea laterali gemina; ventre argenteo.

B 3, D 9, P 13, V 8, A 11, C 18+.

72. *Cyprinus Cosuatis*, Cabdio radiis in pinna dorsali 10, in anali 7; linea laterali solitaria; corpore opaco.

B 3, D 10, P—, V 9, A 7, C—.

73. *Cyprinus Guganio*, Cabdio radiis in pinna dorsali 10, in anali 7; linea laterali obsoleta; corpore diaphano.

B 4, D 10, P 12? V 9, A 7, C—.

Species 74. *Cyprinus Cotio*, Cabdio radiis in pinna dorsali 10, in anali 36, in ventralibus 12.

B 3, D 10, P 16, V 12, A 36, C 19+.

75. *Cyprinus Devario*, Cabdio radiis in pinna dorsali 18, in anali totidem.

B 3, D 18, P 10, V 8, A 18, C 19+.

76. *Cyprinus Dancena*, Cabdio radiis pinnæ dorsalis 6, analis 20.

B—, D 6, P 10, V 6, A 20, C 16.

Subgenus IX. Garra Cyprini absque ulla ad aliud genus affinitate; corpore parvo, vix compressiusculo; absque maculis, vittis, notave colorum alia insigni.

77. *Cyprinus Lamta*, Garra cirrhis 4; pinnis pectoralibus medio acuminatis; radiis pinnæ dorsalis 10, analis 7; linea laterali supera.

B 3, D 10, P 13, V 9, A 7, C 19+.

78. *Cyprinus Sada*, Garra cirrhis 4; pinnis pectoralibus supra acuminatis; radiis pinnæ dorsalis 10, analis 7; linea laterali supera.

B 3, D 10, P—, V 9, A 7, C—.

79. *Cyprinus Latius*, Garra cirrhis duobus; radiis in pinna dorsali 11, in anali 7.

B 3, D 11, P 13? V 9, A 7, C 20+.

80. *Cyprinus Gohama*, Garra? cirrhis duobus; radiis in pinna dorsali 11, in anali 8.

B 3, D 11, P 13? V 9, A 8, C 20+.

81. *Cyprinus Mosario*, Garra imberbis pinnis pectoralibus supra acuminatis; lateribus nebulorum expertibus; radiis pinnæ dorsalis 10, analis 7.

B 3, D 10, P 13? V 9, A 7, C 20.

82. *Cyprinus Sucatio*, Garra imberbis pinnis pectoralibus rotundatis; lateribus nebulosis, radiis in pinna dorsali 9, in anali 7.

B 0? D 9, P 13, V 9, A 7, C 16+.

Species 83. *Cyprinus Balitora,* Garra imberbis pinnis pectoralibus rotundatis; maculis nebulosis triplice serie dispositis, una in dorso, altera in latere utroque; radiis pinnæ dorsalis 10, analis 7.

B 2, D 10, P 12? V 9, A 7, C 18+.

Genus XXI. Cobitis pinna dorsali unica radiata; radiis dorsalibus pectoralibusque muticis; maxillis edentulis; cirrhis vel 6 vel 8.

Species 1. *Cobitis Botia,* oculis subtus armatis; cirrhis 6; corpore nebuloso; radiis in pinna dorsali 14, in ventralibus 8.

B 3? D 14, P 13? V 8, A 8, C 15+.

2. *Cobitis Gongota,* oculis subtis armatis; cirrhis 6; corpore nebuloso; radiis in pinna dorsali 11, in ventralibus 6.

B 0, D 11, P 10, V 6, A 8, C 17.

3. *Cobitis Cucura,* oculis subtus armatis; cirrhis 6; corpore nebuloso; radiis in pinna dorsali 9, in ventralibus 6.

B 0, D 9, P 9? V 6, A 7, C 15.

4. *Cobitis Guntea,* oculis subtus armatis; cirrhis 6; corpore nebuloso; radiis in pinna dorsali 8, in ventralibus 7.

B 3? D 8, P 8, V 7, A 7, C 17.

5. *Cobitis Dario,* oculis subtus armatis; cirrhis 6; corpore fasciato.

B 0, D 11, P 13, V 8, A 7, C 18.

6. *Cobitis Pangia,* oculis subtus armatis; cirrhis 8; corpore immaculato.

B 0, D 7? P 10? V 6, A 7? C 18?

7. *Cobitis Geto,* oculis subtus armatis; cirrhis 8; corpore fasciato.

B 3, D 12, P 13, V 9, A 7, C—.

8. *Cobitis Balgara,* oculis subtus armatis; cirrhis 8; corpore maculato.

B 0? D 7, P 7, V 7, A 7, C 17.

9. *Cobitis Savona,* oculis inermibus; cirrhis 6; corpore semifasciato.

B—, D 10, P 10? V 6? A 6, C—.

Species 10. *Cobitis Turio*, oculis inermibus; cirrhis 6; corpore nebuloso; radiis in pinna dorsali 8, in ventrali totidem.

B 0? D 8, P 12? V 8, A 7, C 19.

11. *Cobitis Bilturio*, oculis inermibus; cirrhis 6; corpore nebuloso; radiis in pinna dorsali 14, in ventralibus 8.

B 0? D 14, P 14, V 8, A 7, C—.

12. *Cobitis Corica*, oculis inermibus; cirrhis 6; corpore maculato; radiis in pinna dorsali 9.

B—? D 9, P 11, V 7, A 6, C 19+.

INDEX.

FINIS.